Optimizing Wireless/RF Circuits

Optimizing Wireless/RF Circuits

John D. Lenk

McGraw-Hill

New York San Francisco Washington, D.C. Auckland Bogotá
Caracas Lisbon London Madrid Mexico City Milan
Montreal New Delhi San Juan Singapore
Sydney Tokyo Toronto

Library of Congress Cataloging-in-Publication Data

Lenk, John D.
    Optimizing wireless/RF circuits / John D. Lenk.
       p.   cm.
    Includes index.
    ISBN 0-07-135226-0
    1. Radio circuits. 2. Electronic circuit design. I. Title.
    TK6560.L42   1999
    621.384'12—dc21   99-35697
                                                                                  CIP

McGraw-Hill

A Division of The McGraw·Hill Companies

Copyright © 1999 by The McGraw-Hill Companies, Inc. Printed in the United States of America. Except as permitted under the United States Copyright Act of 1976, no part of this publication may be reproduced or distributed in any form or by any means, or stored in a data base or retrieval system, without the prior written permission of the publisher.

1 2 3 4 5 6 7 8 9 0 DOC/DOC 9 0 9 8 7 6 5 4 3 2 1 0 9

ISBN 0-07-134376-8

The sponsoring editor of this book was Scott Grillo. The editing supervisor was Peggy Lamb, and the production supervisor was Pamela Pelton. This book was set in New Century Schoolbook per the MHT design by Paul Scozzari of McGraw-Hill's Professional Book Group composition unit, Hightstown, N.J.

Printed and bound by R. R. Donnelley & Sons Company.

This book is printed on recycled, acid-free paper containing a minimum of 50% recycled, de-inked fiber.

LIMITS OF LIABILITY AND DISCLAIMER OF WARRANTY
The authors and publisher have exercised care in preparing this book and the programs contained in it. They make no representation, however, that the programs are error-free or suitable for every application to which the reader may attempt to apply them. The authors and publisher make no warranty of any kind, expressed or implied, including the warranties of merchantability or fitness for a particular purpose, with regard to these programs or the documentation or theory contained in this book, all of which are provided "as is." The authors and publisher shall not be liable for damages in amount greater than the purchase price of this book, or in any event for incidental or consequential damages in connection with, or arising out of the furnishing, performance, or use of these programs or the associated descriptions or discussions.
Readers should test any program on their own systems and compare results with those presented in this book. They should then construct their own test programs to verify that they fully understand the requisite calling conventions and data formats for each of the programs. Then they should test the specific application thoroughly.

Greetings from the Villa Buttercup!
To my wonderful wife, Irene.
Thank you for being by my side all these years!
To my lovely family, Karen, Tom, Brandon, Justin and Michael
And to our Lambie and Suzzie, he happy wherever you are!
To my special readers; May good fortune find your doorway,
bringing you good health and happy things.
Thank you for buying my books!
Special thanks to Scott Grillo, Steve Chapman, Ted Nardin,
Mike Hays, Patrick Hansard, Peter Mellis,
Andrew Yoder (best-selling author), and Robert McGraw of
McGraw-Hill for making me an international best seller again!
This is book number 94
Abundance!

CONTENTS

Preface   xi
Acknowledgments   xiii

Chapter 1. RF Basics   1

1.1   RF Bands   1
1.2   RF Circuit Types   1
1.3   Narrowband RF   3
1.4   RF Power-Amplifier Circuits   6
1.5   Wideband RF   8
1.6   IC RF Voltage Amplifiers   12

Chapter 2. Practical Considerations for RF   25

2.1   Resonant Circuits   25
2.2   Resonant Frequency versus Q or Selectivity   27
2.3   Calculating Resonant Values   28
2.4   Inductance of RF Coils   29
2.5   Voltage-Variable Capacitors (VVCs)   29
2.6   Basic RF-Circuit Design Approaches   30
2.7   The y parameter System   31
2.8   The Four Basic y Parameters   32
2.9   Measuring y Parameters   34
2.10   Using the Smith Chart   38
2.11   RF-Amplifier Stability   41
2.12   RF-Amplifier Stability Solutions   43
2.13   RF Circuits with Neutralization   45
2.14   RF Circuits without Neutralization (Mismatching)   46
2.15   Large-Signal Design Approach   46
2.16   Discrete-Component RF-Oscillator Circuits Design   57
2.17   Thermal Design Considerations   65
2.18   RF Test Equipment   68
2.19   Basic RF Measurements   77

Chapter 3. Direct-Conversion Tuners for Digital DBS — 85

   3.1 Circuit Description for MAX2102 — 85
   3.2 Front-End Tuner Circuitry For DBS Tuners — 87
   3.3 External Oscillator Requirements — 89
   3.4 Prescaler Requirements — 89
   3.5 Baseband Amplifiers — 90
   3.6 Offset Correction — 90
   3.7 Optimizing MAX2102 Layout — 91
   3.8 Power-Supply Sequencing — 91
   3.9 Low-Pass Filters in Direct-Conversion Tuners — 91

Chapter 4. IF Transceiver with Limiter and RSSI — 95

   4.1 Circuit Description for MAX2511 — 95
   4.2 Receiver Circuit — 96
   4.3 Transmitter Circuit — 99
   4.4 Local Oscillator and Oscillator Buffer — 100
   4.5 Mode Selection — 100
   4.6 200 to 440-MHz RF Applications — 101
   4.7 Oscillator Tank Calculations — 101
   4.8 Impedance Matching for the Receiver Input — 102
   4.9 Filter Sharing — 103
   4.10 Receiver IF Filter — 104
   4.11 Optimizing MAX2551 Layout — 104

Chapter 5. RF Power Transistors for 900 MHz — 107

   5.1 Transistor Characteristics — 107
   5.2 Current-Mirror Bias — 108
   5.3 Optimum Port Impedance — 109
   5.4 Optimizing MAX2601/2602 Layout — 110

Chapter 6. IC Oscillator for 650 to 1050 MHz — 111

   6.1 MAX2620 Characteristics — 111
   6.2 Oscillator Circuit — 111
   6.3 Output Buffers — 112
   6.4 Oscillator Tank-Circuit Design — 113
   6.5 Matching the Output Buffers — 114

Chapter 7. General-Purpose Amplifiers for VHF to Microwave — 117

   7.1 MAX2630 through 2633 Characteristics — 117
   7.2 External Components for Typical Operating Circuits — 118
   7.3 Optimizing MAX2630 through MAX2633 Layout — 119

Chapter 8. Applications for Special-Purpose RF ICs — 123

   8.1 Wideband VHF Antenna Booster — 123

8.2	Tuned Amplifier (Matching for Increased Gain)	124
8.3	High-Performance Mixer	127
8.4	Double-Conversion PLL Detector and RF Mixer	135

Chapter 9. Optimizing Frequency Synthesizers — 143

9.1	Loop Bandwidth	143
9.2	Multimodulus Division	145
9.3	SP8853 Characteristics	148
9.4	Prescaler and A and M Counters	148
9.5	Reference Source and Divider	149
9.6	Phase Comparator	150
9.7	Data Entry and Storage	151
9.8	Optimizing SP8853 Circuits	154
9.9	Optimizing the Loop Filter	156
9.10	Second-Order Loop Filter Calculations	157
9.11	Third-Order Loop Filter Calculations	159

Chapter 10. Optimizing High-Speed Frequency Dividers — 163

10.1	PC Boards for High-Speed Dividers	163
10.2	Components Used at High Frequencies	163
10.3	Single-Point Grounding	164
10.4	Impedance Matching PC Board Traces	164
10.5	Frequency Range and Input Level	165
10.6	Dividers with ECL Outputs	166
10.7	Dividers with Open-Collector TTL Outputs	166
10.8	Dividers with True TTL Outputs	166
10.9	Dividers with CMOS Outputs	166
10.10	ECL-TTL Interface	167
10.11	Interfacing High-Speed Dividers	167
10.12	Matching the Input Impedance of High-Speed Dividers	167
10.13	Optimizing Variable Modulus Dividers	168
10.14	Using Dividers in Microwave Synthesizers	172
10.15	Using Dividers in VHF Synthesizers	172

Chapter 11. Optimizing Frequency Synthesizer Design — 175

11.1	Basic Single-Loop FS PLL	175
11.2	Addressing in the Self-Programming Internal Mode	175
11.3	Addressing in the Single-Shot Internal Mode	177
11.4	Addressing in the External Mode	177
11.5	Frequency Synthesizer IC Pin Functions	177
11.6	Programming Considerations	179
11.7	Phase Comparators	179
11.8	Reference Divider Programming	180
11.9	A and M Divider Programming	181
11.10	Calculator Program for Synthesizer Programming	182

x Contents

Chapter 12. Optimizing Direct Frequency Synthesizers — 185

- 12.1 The Basic DFS Circuit — 185
- 12.2 SP2001 Circuit Description — 186
- 12.3 Optimizing the SP2001 Layout — 190
- 12.4 Minimizing Spurious Outputs — 192

Chapter 13. Optimizing Universal Radio ICs — 195

- 13.1 SL6700 Circuit Description — 195
- 13.2 SL6700 as a Double-Conversion IF Strip — 196
- 13.3 SL6700 as an AM Broadcast Radio — 197
- 13.4 SL6700 as an AM/SSB/CW IF Strip — 199
- 13.5 SL6700 as an SSB Generator — 201
- 13.6 SL6700 as a Remote-Control Receiver — 204

Chapter 14. Optimizing Log/Linear Amplifier ICs — 207

- 14.1 SL3522 Circuit Description — 207
- 14.2 Optimizing the RF-Output Buffer — 209
- 14.3 Optimizing SL3522 Circuits — 210
- 14.4 Optimizing Video Performance — 212
- 14.5 Gain and Offset Trimming — 212

Index   213

Preface

This book is a "crash course" in wireless/RF circuit technology, with something for everyone involved in electronics. No matter what your skill level, the book puts you on top of the wireless/RF picture quickly, and then goes on to fill in the details.

For the design engineer and experimenter, the book provides sufficient information to design and build wireless/RF circuits from scratch. The design approach here is the same as that used in all of the author's best-selling books on simplified and practical design.

Throughout the book, design problems start with guidelines for selecting components and ICs on a trial-value basis, assuming a specific design goal and set of conditions. Then, using the guideline values in experimental form, the desired results (frequency range, power output, intermodulation, phase noise, dynamic range, etc.) are produced by varying the experimental component values, as needed, for optimum performance.

If you are an engineer responsible for designing wireless/RF circuits or selecting ICs, the variety of circuit configurations described here will generally simplify your task. Because most present-day wireless/RF technology is in IC form, the book concentrates on such devices, covering a cross-section of the most-popular RF ICs available.

Although this wealth of information permits you to find the right IC for your particular circuit need, the book does not stop there. Each IC features a full description of how the internal circuit operates, how these operational functions affect circuit performance and, of greatest importance, how to select external components to get the best possible results for a particular design.

The first two chapters provide a review and summary of RF basics from a practical standpoint. This information (such as using y-parameters and Smith charts, RF-circuit stability calculations, and basic large-signal RF design) is included for those who are not completely familiar with wireless/RF technology, and for those who need a quick refresher.

The descriptions included in these chapters form the basis for understanding operation of the many ICs covered in the remaining chapters.

The preliminary chapters conclude by describing practical considerations for RF (such as resonant-circuit design and thermal/mounting problems, and PC board layout).

The final chapters are devoted to a cross-section of wireless/RF ICs. Such ICs are used in a variety of applications, including direct-broadcast satellite (DBS) tuners, AM/SSB receivers, log/limiting amplifiers, frequency synthesizers, cellular phones, wireless local loops, broadband systems, wireless handsets, IF transceivers, wireless data links, two-way pagers, land-mobile radios, global positioning systems (GPS), TV tuners, set-top cable boxes, and cordless phones. In each case, the IC is described fully from a practical design standpoint, with special emphasis on selecting external components and board layout for optimum performance in a specific application.

Acknowledgments

Many professionals have contributed to this book. I gratefully acknowledge the tremendous effort needed to produce this book. Such a comprehensive work is impossible for one person, and I thank all who contributed, both directly and indirectly.

I give special thanks to the following: Alan Haun of Analog Devices, Syd Coppersmith of Dallas Semiconductor, Rosie Hinojosa of EXAR Corporation, Jeff Salter of GEC Plessey, John Allen, Helen Cox, and Linda da Costa of Harris Semiconductors, Ron Denchfield and Bob Scott of Linear Technology Corporation, David Fullagar and William Levin of Maxim Integrated Products, Fred Swymer of Microsemi Corporation, Linda Capcara of Motorola, Inc., Andrew Jenkins and Shantha Natarajan of National Semiconductor, Antonio Ortiz of Optical Electronics Incorporated, Lawrence Fogel of Philips Semiconductors, John Marlow of Raytheon Electronics Semiconductor Division, Anthony Armstrong of Semtech Corporation, Ed Oxner and Robert Decker of Siliconix Incorporated, Amy Sullivan of Texas Instruments, Alan Campbell of Unitrode Corporation, Sally and Barry E. Brown (Broker), and Andrew Yoder (best-selling author).

I also wish to thank Joseph A. Labok of Los Angeles Valley College for help and encouragement throughout the years.

And a very special thanks to Steve Chapman, Scott Grillo, Stephen Fitzgerald, Leslie Wenger, Patric Hansard, Peter Mellis, Ted Nardin, Mike Hays, Lisa Schrager, Mary Murray, Carol Wilson, Judy Kessler, Monika Macezinskas, Florence Trimble, Fran Minerva, Jane Stark, Fred Perkins, Robert McGraw, Judith Reiss, Charles Love, Betty Crawford, Jeanne Myers, Peggy Lamb, Thomas Kowalczyk, Clare Stanley, Suzanne Rapcavage, Jaclyn Boone, Kathy Green, Donna Namorato, Regina Frappolli, Pamela Pelton, Sherri Souffrance, Allison Arias, and Midge Haramis of the McGraw-Hill Professional Publishing organization for having that much confidence in me.

And to Irene, my wife and Super Agent, I extend my thanks. Without her help, this book could not have been written.

Chapter 1

RF Basics

This chapter is devoted to a review and summary of RF basics from a simplified design standpoint. The information here is provided for those readers who are totally unfamiliar with RF (or those who think they know it all).

1.1 RF Bands

Electric signals (passing through a conductor) generate *electromagnetic waves*, which are radiated or transmitted from the conductor when the signal frequency is about 15 kHz and higher. Because of this radiating property, signals of such frequencies are known as *radio-frequency (RF) signals*. It is not practical to design any circuit that covers the entire frequency range or to use all radio frequencies for all purposes. Instead, the RF spectrum is broken down into various bands, each used for a specific purpose. In turn, RF circuits are generally designed for use in one particular band.

Figure 1.1 shows the most common assignment of RF bands, including both commercial and military bands. Notice that radio waves at frequencies above about 1 GHz are known as *microwaves*; the circuits used with microwaves are quite different from those used at lower frequencies. Because of their specialized nature, microwave circuits are not covered in any detail. Instead, this book concentrates on RF circuits operating at frequencies up to and including the UHF band.

1.2 RF Circuit Types

The most common type of RF circuit is the amplifier, with the oscillator (covered in Chap. 2) running a close second. RF amplifiers are often

RF-Band Frequency Ranges

Commercial Bands

Very low frequency (VLF) 3–30 kHz
Low frequency (LF) 30–300 kHz
Medium frequency (MF) 300 kHz–3 MHz
High frequency (HF) 3–30 MHz
Very high frequency (VHF) 30–300 MHz
Ultrahigh frequency (UHF) 300 MHz–3 GHz
Superhigh frequency (SHF) 3–30 GHz
Extrahigh frequency (EHF) 30–300 GHz

Military Bands

P-band 225–390 MHz
L-band 390–1550 MHz
S-band 1.5–5.2 GHz
X-band 5.2–10.9 GHz
K-band 10.9–36 GHz
Q-band 36–46 GHz
V-band 46–56 GHz

United States Broadcast Bands

Amplitude modulated (AM) 535–1605 kHz
Frequency modulated (FM) 88–108 MHz
VHF television 54–216 MHz
UHF television 470–890 MHz

Figure 1.1 Common assignment of RF bands.

divided into two general types: *narrowband amplifiers* (with bandwidths up to several hundred kHz) and *wideband amplifiers* (with bandwidths of several MHz). The following describes the need for both types of amplifiers.

As shown in Fig. 1.1, the AM broadcast band for the United States runs from 540 to 1700 kHz. The frequencies of broadcast stations within this band are spaced 10 kHz apart to prevent interference with each other. In the FM broadcast band, the transmitting-station frequencies are spaced 200 kHz apart. In the nondigital TV broadcast bands, the stations are spaced about 6 MHz apart.

Within a specific band, each transmitting station is assigned a specific frequency at which the station is to operate. However, each station also transmits across a narrow band of frequencies at either side of the assigned frequency. The band of frequencies is required if the signal is to convey audio/video intelligence. For example, an AM broadcast station transmits with a band extending 5 to 7.5 kHz on either side of the assigned frequency.

An RF amplifier used in an AM broadcast radio receiver is adjusted to cover a portion of the band about 15 kHz wide, corresponding to the

15-kHz spread of a single station. Under these conditions, the *bandwidth* of the RF amplifier is said to be 15 kHz. The amplifier is adjusted (tuned) to one station at a time. In the FM broadcast band, where each station is spaced 200 kHz apart, the bandwidth of the RF amplifier is about 150 kHz. Both AM and FM broadcast-band RF amplifiers are essentially narrowband amplifiers. In the television bands, where the station channels are 6 MHz apart, the RF amplifiers are of the wideband type (or *broadband* type) because the transmitted RF signal is about 4.5 MHz wide.

To sum up, an RF amplifier serves two purposes. One is as a *bandpass filter*, which passes signals from the desired station and rejects all others; the other is to amplify these signals to a suitable voltage (power) level.

1.3 Narrowband RF

Figure 1.2 shows a typical narrowband RF amplifier, such as that used in older discrete-component radio receivers (the typical transistor radio). In present design, the circuit is usually part of an IC tuner package, as described in Sec. 1.6. For now, consider the bandpass filter or tuning function and certain feedback problems.

1.3.1 Narrowband RF tuning

The circuit of Fig. 1.2 is a single stage of tuned radio frequency (TRF) voltage amplification. The input to the stage is by means of transformer

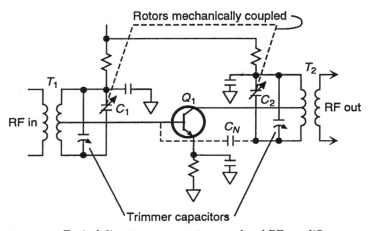

Figure 1.2 Typical discrete-component narrowband RF amplifier.

T1; the output is taken from transformer T2. The secondary of T1 is tuned to resonance (Chapter 2) at the frequency of the incoming signal by variable capacitor C1; T2 is tuned to the same resonant frequency by C2.

In many cases, the transformers are tuned by adjustable powdered-iron cores, with fixed capacitors for C1 and C2. Because it is impossible to build two RF circuits that are exactly alike, trimmer capacitors (in parallel with C1 and C2) vary the overall capacitance of the tuned circuit slightly to compensate for small differences between the circuits.

Although the use of several tuned circuits increases the overall *selectivity* of the amplifier, the need for manipulating a number of variable capacitors can be a problem. This situation is overcome by *ganging,* in which the rotors of variable capacitors are mechanically connected so that all rotors move simultaneously when one dial is turned. Any variations in tuning are corrected by the trimmers on each transformer. When the transformers are tuned by adjustable cores, ganging is not used and each tuning circuit is adjusted separately.

1.3.2 RF feedback problems

Undesired feedback (particularly from output to input) is a problem with all RF circuits. The two types of such feedback are: radiated feedback and feedback through the transistor (internal feedback). Radiated feedback is prevented by shielding. Fortunately, most modern transistors show little internal feedback at lower frequencies, but feedback can be a problem when frequency increases.

Miller effect. The most common RF feedback is known as *Miller effect* and is caused by input and output capacitances of the transistor. Figure 1.3 shows the relationships. There is some capacitance between the base and emitter of a bipolar (or two-junction) transistor, or between the gate and source of a field-effect transistor (FET). These form the input capacitance of an RF circuit. A capacitance is also between the base and collector (gate and drain). This capacitance feeds back some of the collector signal to the base.

The collector signal is amplified and is 180° out of phase with the base signal (in a common-emitter amplifier, such as that shown in Fig. 1.2). The collector signal feedback opposes the base signal and tends to distort the input signal. Also, the collector-base capacitance is, in effect, in series with the base-emitter capacitance. Thus, it changes the input capacitance.

All these conditions make for a constantly changing relationship of signals in an RF circuit. For example, if the input-signal amplitude

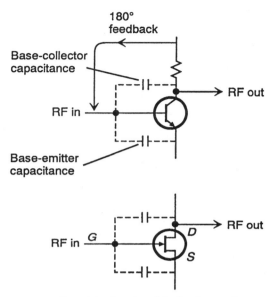

Figure 1.3 Transistor input and feedback capacitances.

changes, the amount of feedback changes, changing the input capacitance. In turn, the change in input capacitance changes the match between the transistor and the input-tuned circuit, changing the amplitude. Also, if the input-signal frequency changes, the feedback changes (because the collector-base capacitive reactance changes), and there is a corresponding change in amplification.

This Miller effect is not necessarily a problem in all RF circuits. FET RF circuits are usually more susceptible to the Miller effect than bipolar transistor circuits. However, when the Miller effect becomes severe with any RF circuit, the effect can be eliminated (or minimized to a realistic level) by neutralization.

Neutralization. With neutralization, a portion of the signal from the output of the RF circuit is fed back to the input so as to cancel any signal produced by unwanted feedback. This can be used to reduce unwanted feedback resulting from both radiated and internal feedback. Neutralization is performed by applying a signal to the input that is equal in magnitude, but opposite in phase, to the undesired feedback. The two signals then cancel each other.

The two ends of an output-transformer primary winding (such as T2 in Fig. 1.2) are of opposite phase. If the opposite-phase signal is fed to the input through a neutralizing capacitor (C_N in Fig. 1.2), the two signals cancel out. As a guideline, the neutralizing capacitor is typically equal to the collector-base capacitance (often a few picofarads).

Common-base RF circuits. Figure 1.4 shows a typical common-base RF amplifier. Such circuits are sometimes used in RF work to reduce unwanted feedback, but without neutralization. Again, the input transformer (T1) is tuned to resonance by variable C1; the output transformer (T2) is tuned to the same resonant frequency by C2. The base is grounded and the input signal is applied to the emitter. The output is taken between the collector and base, which is common to the input and output circuits.

The grounded base acts as a shield between the input and output circuits, thus reducing feedback. The circuit of Fig. 1.4 is often found at the antenna input of discrete-component radio receivers. In addition to minimizing undesired feedback, the ground (common-base) circuit provides a low input impedance (to match a 50-, 75-, or 300-ohm antenna).

1.4 RF Power-Amplifier Circuits

Figure 1.5 shows two basic RF power-amplifier circuits. Most radio transmitters (CB, amateur, business radio, cellular telephones) use some form of power amplifier to raise the low-amplitude signal developed by the oscillator to a high-amplitude signal that is suitable for transmission. For example, most oscillators develop signals of less than 1 W, whereas a transmitter might require an output of several hundred watts (or more).

In the circuit of Fig. 1.5A, the collector load is a parallel-resonant circuit (called a *tank circuit*), consisting of variable capacitor C1 and coil L1, tuned to resonance at the desired frequency. The output, which is an amplified version of the input voltage, is from L2, which (together with L1) forms an output transformer.

The circuit of Fig. 1.5A has certain advantages and disadvantages. The winding of L2 can be made to match the impedance of the load (by

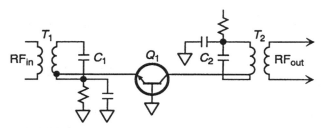

Figure 1.4 Typical discrete-component common-base RF amplifier.

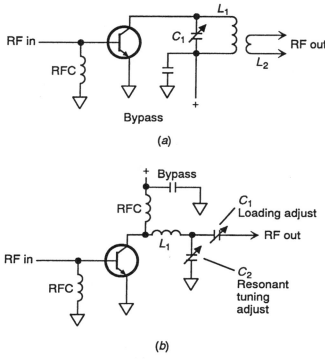

Figure 1.5 Basic RF power-amplifier circuits.

selecting the proper number of turns and positioning L2 in relation to L1). Although that might prove to be an advantage in some cases, it also makes for an *interstage coupling network* that is subject to mismatch and detuning by physical movement or shock. Another disadvantage of the Fig. 1.5A circuit is that all current must pass through the tank-circuit coil.

For best transfer of power, the impedance of L1 should match that of the transistor output. Because bipolar-transistor output impedances are generally low, the value of L1 must be low, often resulting in an impractical size for L1. The circuit of Fig. 1.5A is a carryover from vacuum-tube circuits and, as such, is not often found in present-day RF equipment. A possible exception is in the few low-power FET amplifier circuits.

The circuit of Fig. 1.5B, or one of the many variations, is commonly found in RF equipment using bipolar transistors. The collector load is a resonant circuit formed by the network L1, C1, and C2. Notice that C1 is labeled "Loading adjust," whereas C2 is labeled "Resonant tuning adjust."

As covered in Chapter 2, these networks provide the dual function of *frequency selection* (equivalent to the tank circuit) and *impedance matching* between transistor and load. To properly match impedances, both the resistive (so-called *real part*) and reactive (so-called *imaginary part*) components of the impedance must be considered.

1.4.1 RF multiplier

The circuits in Fig. 1.5 can be used as a frequency multiplier where the collector is tuned to a higher whole-number multiple (harmonic) of the input frequency. Many radio transmitters use some form of multiplier to raise the low-frequency signal developed by the oscillator to a high frequency. Although the circuits in RF power multipliers and RF power amplifiers are essentially the same, the efficiency is different. That is, an RF amplifier operating at the same frequency as the input has a higher efficiency than an amplifier operating at a multiple of the input frequency.

1.4.2 RF amplifier-multiplier combinations

The circuits of Fig. 1.5 can be *cascaded* (the output of one circuit applied to the input of the next circuit) to provide increased power amplification and/or frequency multiplication. Typically, no more than three stages are so cascaded. The stages can be mixed. That is, one or two stages can provide frequency multiplication, with the remaining one or two stages providing power amplification.

1.5 Wideband RF

Except for pure sine waves, all signals contain not only the fundamental frequency, but also harmonic and subharmonic frequencies as well. These harmonics are whole-number multiples of the fundamental frequency. Pulse signals have an especially high harmonic content.

A typical RF amplifier with a bandwidth of several hundred kilohertz is not able to uniformly amplify signals with such a broad range of frequencies. For this reason, it is necessary to use special *broadband* or *wideband* RF amplifiers for such applications. These amplifiers are usually known as *video-frequency (VF) amplifiers* (or simply *video amplifiers*) in television equipment (TV sets, camcorders, VCRs, etc.) or as *pulse amplifiers* in radar and similar equipment (even though they are RF).

Resistance-coupled (RC) amplifiers are, in effect, wideband amplifiers. Such circuits amplify uniformly at all frequencies of the audio range, dropping off only at the low- and high-frequency ends. A wideband amplifier (capable of passing RF signals, including pulses) is formed when the uniform amplification is extended to both ends of the frequency range.

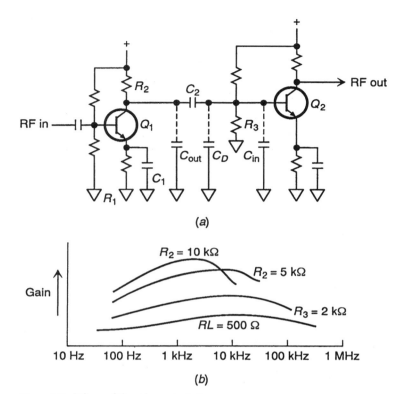

Figure 1.6 RC amplifier characteristics.

Figure 1.6 shows a basic RC amplifier circuit. Capacitance C_{out} represents the output capacitance of Q1, and capacitance C_D represents the *distributed capacitance* of the various components and related wiring. Capacitance C_{in} represents the input capacitance of Q2.

Coupling C2 and base resistor R1 form a voltage divider across the input of Q2. At low frequencies, the impedance of C2 is large, and relatively little of the signal is applied to the base of Q2. Accordingly, the low-frequency response of the circuit is lowered.

Capacitances C_{out}, C_D, and C_{in}, acting in parallel, shunt the load resistor (R2) of Q1. This lowers the effective resistance of R2, as well as the high-frequency response of the circuit. (A lower value of R_2 lowers the gain, all other factors being equal.)

1.5.1 Increasing wideband response

Several methods are used to improve the low- and high-frequency response of RF wideband amplifiers (or RC amplifiers designed for wideband use). In all cases, transistors with small input and output capacitances are used. Likewise, components are carefully placed so that leads and distributed capacitance are kept to a minimum. The following is a summary of additional methods to increase wideband response.

Collector resistance. The value of collector-load resistor R2 affects the frequency response and gain of the amplifier. The graph in Fig. 1.6 shows the effects produced by various values of R_2. A large R_2 produces a high gain at the middle frequencies, and a steep drop in gain at the high and low frequencies. A small R_2 produces much smaller overall gain, but the proportional drop in gain at the high and low frequencies is also much less than for the larger collector resistances. With the small R_2, gain is uniform over a much wider range of frequencies. In effect, the circuit sacrifices gain for bandwidth. Because of these conditions, wideband RF circuits use low collector resistances and transistors with high gain.

Emitter bypass. The emitter-bypass capacitor C1 in Fig. 1.6 affects the low-frequency gain. The impedance of C1 is higher at lower frequencies. Thus, the circuit gain is lower at lower frequencies. Accordingly, capacitance C_1 must be large enough to offer a low impedance (with respect to R1) at the lowest frequency to be passed.

Coupling capacitance. At low frequencies, the effects of the transistor input/output capacitance and the distributed capacitances are negligible, but the impedance of the coupling capacitor becomes increasingly important. Figure 1.7 shows one method used to compensate for the effects of the coupling capacitor. This circuit is the video amplifier of a typical discrete-component TV set. In present design, all or part of the circuits in Fig. 1.7 are in IC form.

In the Fig. 1.7 circuit, the load resistance for Q1 consists of two parts, R5 and R6, connected in series. Capacitor C3 is the bypass capacitor for R6. At the higher frequencies, R5 is, in effect, the collector load because the small impedance of C2 at these frequencies permits C3 to completely bypass R6. This removes R6 from the circuit, as far as signals are concerned.

At low frequencies, the impedance of C3 becomes high, and the bypassing effect is greatly reduced. The collector-load resistance then becomes R_5 plus R_6. This greater resistance produces a greater output voltage, thus compensating for any low-frequency drop that results from the increasing impedance of coupling capacitors (such as C5 and C7).

Shunt peaking coil. Because the drop in high-frequency response is caused by the shunting effect of the transistor capacitances (and dis-

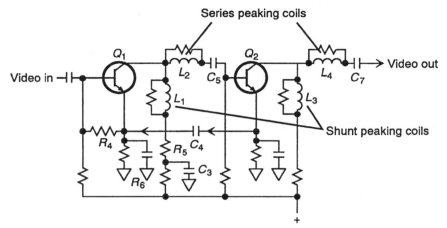

Figure 1.7 Typical discrete-component video amplifier.

tributed capacitances) upon the load resistor, a small coil, L1 (called a *shunt peaking coil*), is inserted in series with the load resistances. At low frequencies, L1 offers very little impedance, and the collector load is, in effect, the resistance of R5 plus R6. At high frequencies, the impedance of L1 is high, and the collector load is the sum of the R5 plus R6 resistances and the resistance of L1. Thus, circuit gain is increased.

Coil L1 sets up a resonant circuit with the distributed capacitances of the circuit (and the capacitances of the transistor). The value of L1 is selected so that the circuit is resonant at a frequency where the high-frequency response of the circuit begins to drop. In this way, an additional boost is given to the gain, and the high-frequency end of the response curve is flattened.

Series-peaking coil. Figure 1.7 shows another similar frequency-compensation circuit. Series-peaking coil L2 is connected in series with coupling capacitor C5. At high frequencies, L2 forms a low-impedance series-resonant circuit with the capacitances, causing a larger signal to appear at the base of Q2.

RF circuits used in video and pulse applications often use both shunt- and series-peaking coils for high-frequency compensation. Typically, the values of these coils are such that resonance is obtained at the highest desired frequency. That is, the capacitances are calculated (or measured), and a corresponding value of inductance is chosen for resonance at the high-frequency end.

Damping resistances. Notice that the coils in Fig. 1.7 are shown with resistances connected in parallel. As covered in Chapter 2, when resistances are connected across coils, the resonant point of the coils is flattened or broadened. Such resistances are often known as *damping resistances*.

Inverse feedback. Another method to overcome the effects of the drop in gain at the low and high ends of the frequency band involves the use of inverse feedback, provided by C4 and R4 in Fig. 1.7. This feedback (also known as *negative feedback*) opposes changes in signal level. For example, if the base of Q1 is swinging positive, the collector of Q1 swings negative, as do the base and emitter of Q2. The negative Q2 emitter signal is fed back to the Q1 base through C4 and R4, thus opposing the original positive swing. This action occurs at both the high and low ends of the frequency band.

1.6 IC RF Voltage Amplifiers

This section describes the RF-amplifier circuits of a typical AM/FM tuner of IC design, and the IF and video circuits of an IC-based TV set. Although most of the components for both of the circuits are contained in ICs, a number of discrete components are between the ICs, as well as discrete components used to tune or adjust the ICs.

1.6.1 Relationship of AM/FM tuner circuits

Figure 1.8 shows the relationship of the tuner circuits. Figures 1.9 and 1.10 show further details of the FM and AM sections, respectively.

Amplifier and multiplex decoder. IC301 is both an amplifier and a multiplex decoder. It is used in the audio path for AM and FM. When audio reaches pin 2 of IC301, whether from the AM or FM sections, IC301 produces corresponding audio at pins 6 and 7. IC301 has only one adjustment control. This multiplex VCO-adjust pot R305 is connected at pin 15.

FM section. FM broadcast signals are applied to FM tuner package MD101. Notice that MD101 contains RF voltage amplifiers, a local oscillator, a mixer/converter, IF amplifiers, and a detector, all in IC form. If you are not familiar with these functions, read *Lenk's Audio Handbook* (McGraw-Hill, 1991).

The IF output from MD101 is applied to the FM amplifier and detector IC201 through amplifiers Q101/Q102 and ceramic filters MF201/MF202, as shown in Fig. 1.9. This amplifier-filter combination removes any amplitude modulation and passes only signals of the desired frequency. Notice that Q101 and Q102 are connected as a form

RF Basics 13

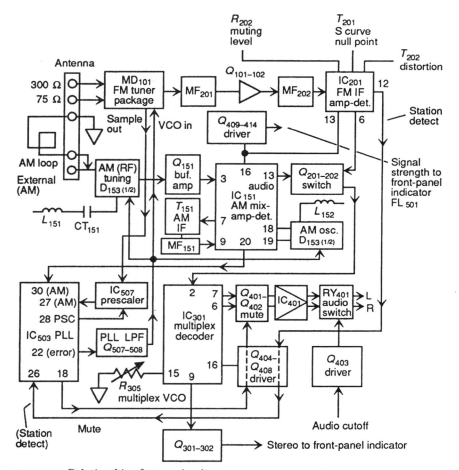

Figure 1.8 Relationship of tuner circuits.

of differential amplifier. Also notice that IC201 has three adjustments: muting-level R202, S-curve null-point T201, and FM distortion T202. The audio output from the FM section is applied to amplifier-multiplex decoder IC301 through audio-select switches Q201 and Q202. These circuits select the AM or FM audio for the input to IC301.

Notice that IC301 has a dual function. When suitable FM is present, IC301 is a stereo amplifier (as well as a decoder). IC301 functions as a mono amplifier when no FM stereo is present or when the stereo signal is too weak to produce proper FM.

AM section. AM broadcast signals are applied to AM IF amplifier-detector IC151 through buffer-amplifier Q151, as shown in Fig. 1.10.

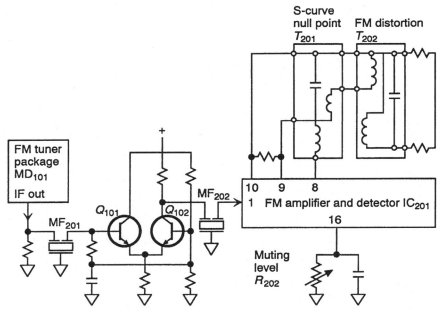

Figure 1.9 FM amplifier circuits in IC tuner.

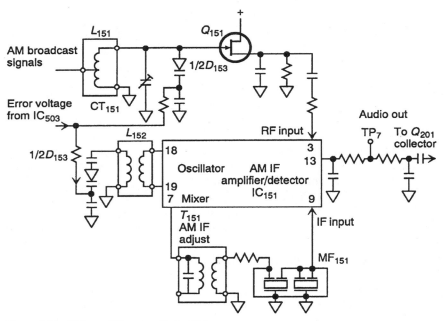

Figure 1.10 AM amplifier circuits in IC tuner.

The RF circuit (L151 and CT151) is tuned by one section of D153 (a voltage-variable capacitor, VVC, such as described in Chapter 2). The mixer output of IC151 is tuned by AM IF adjustment T151 and applied to the IF portion of IC151 through ceramic filter MF151. Notice that IC151 contains RF amplifiers, a local oscillator, a mixer-converter, IF amplifiers, and a detector, similar to that described for the FM section.

The audio output from the detector in IC151 is applied to IC301 through Q201 and Q202. Notice that IC301 functions as a mono amplifier when the AM mode is selected. Because no stereo signal is present in the AM mode, IC301 shifts to mono operation, just as when there is no FM stereo or when the stereo signal is weak.

FM tuning. The FM section is tuned to the desired frequency (and locked to that frequency) by signals applied to MD101, as shown in Fig. 1.8. The VCO signal (a variable dc voltage, sometimes called the *error voltage*) from phase-locked loop (PLL) microprocessor IC503 is applied to MD101 through Q507 and Q508 (which are connected as a direct-coupled amplifier). The IC503 error voltage shifts the MD101 oscillator (as necessary) to tune across the FM broadcast band or to fine-tune MD101 at a selected station.

The frequency produced by MD101 is sampled and applied to the FM input of IC503 through prescaler IC507. (Operation of the prescaler is described in Sec. 1.6.2.) The sampled signal serves to complete the FM tuning loop. For example, if MD101 drifts from the frequency commanded by IC503, the error voltage from IC503 changes the MD101 oscillator (as necessary) to bring MD101 back on frequency.

A station-detect signal is produced at pin 12 of IC201. This signal is used to tell PLL microprocessor IC503 that an FM station has been located and that the station has sufficient strength to produce good FM operation. Between FM stations, or when the FM station is weak, pin 12 of IC201 goes high, turning on Q406. This causes pin 26 of IC503 to go low and causes the audio to be muted (to eliminate background noise when tuning from station to station). When an FM station is of sufficient strength, pin 12 of IC201 goes low, turning Q406 off and causing pin 26 of IC503 to go high. Under these conditions, the audio is unmuted, and IC503 fine-tunes the FM section for best reception of the FM station. The muting level is set by R202.

AM tuning. The AM section is also tuned to the desired frequency (and locked to that frequency) by signals from IC503 applied to the two sections of diode D153. This same error voltage is applied to the VCO of FM tuner MD101. The IC503 error voltage shifts the RF and oscillator circuits (as necessary) to tune across the AM broadcast band.

The oscillator circuit of IC151 is adjusted by L152, as the RF input circuit is tuned by tracking adjustments L151 and CT151. The frequency produced by IC151 is sampled and applied to the AM input of IC503. The sampled signal serves to complete the AM tuning loop. For example, if the AM section drifts from the frequency commanded by IC503, the error voltage from IC503 changes both circuits controlled by D153 (as necessary) to bring the AM section back on frequency.

1.6.2 Frequency synthesis tuning of RF voltage amplifiers

The RF voltage amplifiers in Fig. 1.8 use some form of frequency synthesis tuning (also known as *quartz tuning* or *digital tuning*). Frequency synthesis (FS) tuning provides for convenient pushbutton or preset AM and FM station selection (or TV channel selection, as described in Sec. 1.6.3). Most FS systems also provide automatic station-search or scan, and automatic fine-tune (AFT) capability.

As shown in Fig. 1.11, the key element in an FS system is the PLL that controls the variable-frequency oscillator (VFO) and/or RF tuning, as required for both station selection and fine tuning. Notice that the PLL used in the AM/FM tuner in Fig. 1.8 is essentially the same as the PLL used in the FS tuners of TV sets (Section 1.6.3) and VCRs (and can be applied to virtually any RF tuning system).

Elements of an FS system. As shown in Fig. 1.11A, the basic PLL is a frequency-comparison circuit in which the output of a VPO is compared in frequency and phase to the output of a very stable (usually quartz-crystal controlled) fixed-frequency reference oscillator. If a deviation occurs between the two compared frequencies or if any phase difference is between the two oscillator signals, the PLL detects the degree of frequency or phase error and automatically compensates by tuning the VFO up or down in frequency or phase until both oscillators are locked to the same frequency and phase.

The accuracy and frequency stability of a PLL circuit depend on the accuracy and frequency stability of the reference oscillator (and not the crystal that controls the reference oscillator). Regardless of what reference oscillator is used, the VFO of most PLL circuits is a VCO, in which the frequency is controlled by an error voltage.

AM FS tuning. Figure 1.11B shows how the PLL principles are applied to the AM section of a typical RF circuit (AM/FM tuner). The 1-kHz reference oscillator in Fig. 1.11A is replaced by a reference signal, obtained by dividing down the PLL IC503 clock (4.5 MHz). This reference signal is applied to a phase comparator within IC503. The other input to the phase comparator is a sample of the AM signal at pin 30

RF Basics 17

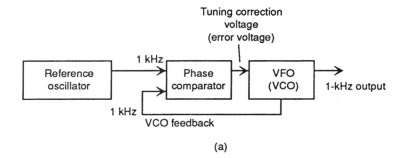

(a)

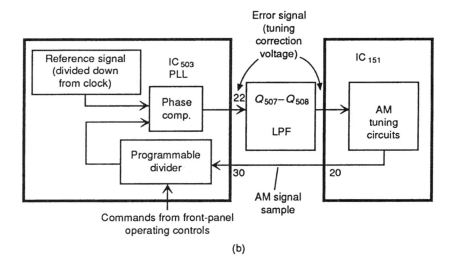

(b)

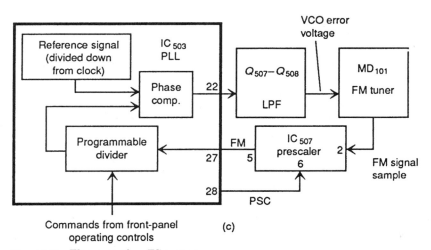

(c)

Figure 1.11 Elements of an FS system.

of IC503 (taken from pin 20 of IC151). The output of the phase comparator is an *error signal* or *tuning-correction voltage* that is applied through low-pass filter Q507 and Q508 to the AM tuning circuits. The filter acts as a buffer between the comparator and tuning circuits.

Notice that the AM sample is applied to the phase comparator through a *programmable divider* or *counter*. The division ratio of the programmable divider is set by commands from the front-panel operating controls (tuning up or down, preset scan, etc.). In effect, the divider is programmed to divide the AM sample by a specific number. The variable-divider function makes possible many AM local-oscillator frequencies across the RF spectrum. An AM frequency change is made by varying the division ratio with front-panel commands. This produces an error signal that shifts the tuning circuits until the AM signal (after division by the programmable divider) equals the reference-signal frequency, and the tuning loop is locked at the desired frequency.

FM FS tuning. Figure 1.11C shows the PLL circuits for the FM section of an RF tuner. This circuit is similar to the PLLs used in TV sets (Sec. 1.6.3) and VCRs in that a *prescaler* is used. The system in Fig. 1.11C is generally called an *extended PLL* and it holds the variable oscillator frequency to some harmonic or subharmonic of the reference oscillator (but with a fixed relationship between the reference and variable signals).

The PLL in Fig. 1.11C uses a form of *pulse-swallow control (PSC)* that allows the division ratio of the programmable divider to be changed in small steps. As in the case of AM, the division ratio of the programmable divider is determined by commands applied to IC503 from the front-panel controls.

The PSC system uses a very high-speed prescaler IC507 and a variable division ratio. The division ratio of the prescaler is determined by the PSC signal at pin 28 of IC503 and can be altered (as required) to produce subtle changes in frequency needed for optimum station tuning (fine tuning) in the FM mode.

The PSC signal at pin 28 of IC503 is a series of pulses. As the number of pulses increases, the division ratio of the prescaler also increases. When a given FM station or frequency is selected by the front-panel controls, the number of pulses on the PSC line is set by circuits in IC503, as necessary for each station or frequency.

The overall division ratio for a specific FM station or frequency is the prescaler division ratio, multiplied by the programmable-divider division ratio. The result of division at any FM station or frequency is a fixed output to IC503 when the FM tuner is set to the desired frequency.

1.6.3 Frequency-synthesis tuning for TV RF circuits

Figures 1.12 and 1.13 show some typical FS tuning circuits in TV sets with multiband RF tuners. These circuits are far more complex than those for AM/FM RF tuners.

The circuit in Fig. 1.12A is similar to the basic PLL circuit in Fig. 1.11A. Figure 1.12B shows a somewhat more sophisticated PLL circuit, one capable of comparing frequencies that are not identical. The circuit in Fig. 1.12B includes a divide-by-10 element, which divides the VCO frequency by 10, and a low-pass filter, which acts as a buffer between the comparator and VCO. Notice that although the inputs to the comparator remain at 1 kHz when the loop is locked, the output frequency of the VCO is now 10 kHz because of the divide-by-10 circuit.

Figure 1.12C shows a PLL circuit that is similar to those found in video equipment (the RF tuners for both TV and VCRs). The system is generally called an *extended PLL*; it holds the tuner-oscillator frequency to some harmonic (or subharmonic) of the reference oscillator. The fixed-divider element in Fig. 1.12B is replaced by a programmable variable divider ($\div N$) in Fig. 1.12C. A channel change is produced by varying the division ratio of the programmable divider with 4-bit data commands from the system-control microprocessor (which, in turn, is operated by front-panel controls and/or by remote-control signals).

TV pulse-swallow control. The RF tuners in most TV sets also use PSC with a variable division ratio, such as that shown in Fig. 1.12D. Here, the variable division ratio depends on the PSC signal from PLL IC1. When the number of PSC pulses increases, the division ratio increases. A specific number of PSC pulses are produced by IC1 in response to channel-selection commands.

The overall division ratio for a specific channel is the prescaler division ratio multiplied by the programmable-divider division ratio. The result of division at any channel is a precise 5-kHz output to the comparator when the tuner oscillator is locked to a given channel frequency. In effect, the programmable divider determines the basic channel frequency and the prescaler performs the fine adjustments (automatic fine tuning or AFT) to the channel frequency.

If you are troubleshooting any RF tuner with PSC and you cannot tune in a channel with AFT (or manually), check the PSC line (pin 27 of IC1, in this case) for pulses and check the tuning voltage to the oscillator. If the pulses are missing, the tuner cannot lock into any channel—even with a tuning voltage present.

Typical TV FS tuning circuit. Figure 1.13 shows the circuits of a typical FS (or digital) RF tuning system with PLL and PSC. Notice that the

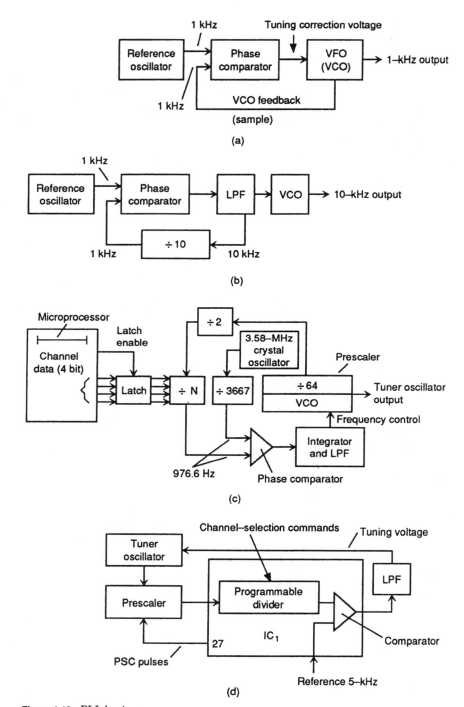

Figure 1.12 PLL basics.

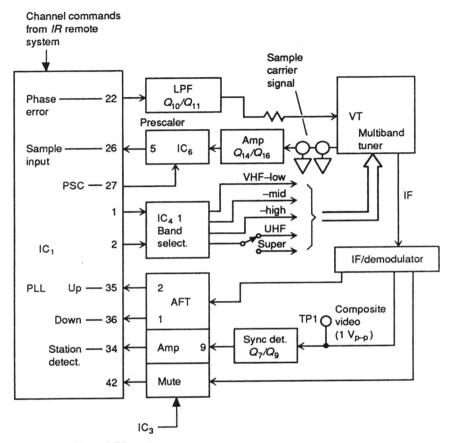

Figure 1.13 Typical FS tuning system.

PLL IC1 receives channel commands from a remote control until after the commands have been decoded by a remote-control receiver unit. The multiband RF tuner (suitable for TV, VCR, and cable) is controlled by circuits within PLL IC1 which, in turn, receives commands from the remote-control circuits. IC1 monitors signals from the IF demodulator circuits to know when a station is being received. These signals are the AFT up and down (pins 35 and 36) and station-detect signal (pin 34).

The detected video from the output of the IF demodulator is passed to a sync amplifier and detector, Q7 and Q9. The detected sync signal (station detect) is passed to pin 9 of IC3, amplified, and applied to the station-detect input at pin 34 of IC1. When a station is tuned in properly, the sync is detected from the video signal and applied as a high to pin 9 of IC3. This applies a high to pin 34 of IC1, indicating to IC1 that video with sync is present.

To maintain proper tuning, IC1 monitors the AFT up and down signals (at pins 35 and 36) from IC3. The AFT circuit of IC3 is a *window detector*, monitoring the AFT voltage and producing a high at pins 1 or 2, depending on the magnitude and direction of the AFT voltage swings (if the tuner-oscillator frequency drifts). When a channel is selected, band-switching information is supplied from IC1 (at pins 1 and 2) to IC4, which develops four band-switching outputs. (In this particular circuit, one of the four outputs is switched by a normal/cable switch to provide the five bands shown.)

The RF tuner oscillator passes a *sample carrier signal* to oscillator amplifier Q14 and Q16. The amplified oscillator signal is then passed to prescaler IC6. The amount of frequency division is determined by PSC pulses from IC1. The frequency-divided output of IC5 (pin 5) is then passed to the sample input of IC1 (at pin 26). When a channel is selected, circuits within IC1 produce the appropriate number of PSC pulses at pin 27. The PSC pulses are applied to IC6 and produce the correct amount of frequency deviation.

The divided-down oscillator signal (sample input) at pin 26 of IC1 is divided down again within IC1 and compared to an internal 5-kHz reference signal. The phase error of these two signals appears at pin 22 of IC1 and is applied to low-pass filter (LPF) Q10 and Q11. The dc output from the LPF is applied to the tuner oscillator. This voltage sets the RF tuner oscillator (as necessary) to achieve the proper frequency for the channel selected.

1.6.4 TV IF/video-detector circuits

Figure 1.14 shows the signal path through the IF and video circuits of a TV set. Although these circuits are called *IF (intermediate frequency) and video,* the signals are at radio frequencies, so the circuits can be considered as RF. The basic function of the circuits in Fig. 1.14 is to amplify both picture and sound signals from the RF tuner, demodulate both signals for application to the video amplifier and sound IF (SIF) amplifiers, and trap (or reject) signals from adjacent RF channels.

In the circuit of Fig. 1.14, IF output from the RF tuner is supplied to an IF amplifier within IC1 through filter CP1 and T1. The IF output is applied to a video detector in IC1. The composite output from the detector is applied through filter CP4 to a video amplifier with IC1. The detector output is also applied to an SIF amplifier (not shown) in IC1.

The output of the video amplifier is applied to the picture-tube circuits through filter CP5, L53, and Q3. The video-amplifier output is also applied to an AGC (automatic gain control) detector within IC1.

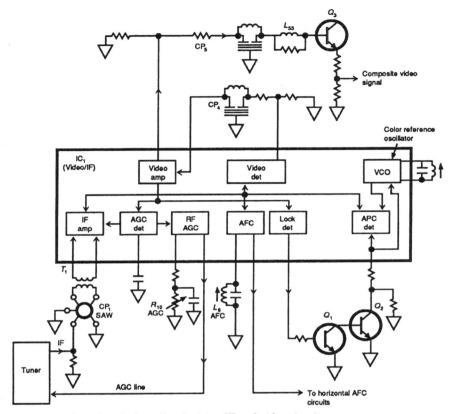

Figure 1.14 Signal path through television IF and video circuits.

This AGC circuit controls both the IF amplifier within IC1 and the tuner (through the RF amplifier in IC1). The AGC circuit is adjusted by R10.

The output of the IC1 video detector is applied to an AFC (automatic frequency control) circuit in IC1. This circuit is adjusted by L6 and provides pulses to the horizontal AFC circuits. The IC1 output is also applied to an APC (automatic phase control) circuit that controls the operation of the color-reference oscillator in IC1 (in conjunction with an IC1 lock detector, which receives signals from the IC1 video amplifier).

Chapter 2

Practical Considerations for RF

This chapter is devoted to practical considerations for optimizing wireless/RF ICs. The chapter covers such subjects as resonant circuits, Q and selectivity, calculating resonant values, RF coils, VVCs, basic RF design approaches, y parameters, Smith charts, RF-amplifier stability, large-signal design approaches, discrete-component RF-oscillator circuit design, thermal design considerations, basic RF testing and measurement techniques, including spectrum analysis, VSWR, resonant frequency measurements, coil inductance, circuit Q and self-resonance, and circuit impedance. As in the case of Chapter 1, this information is included for those readers who need a review and summary of RF basics. The information here forms a basis for understanding the IC optimizing techniques found in the remaining chapters.

2.1 Resonant Circuits

Both RF amplifiers and RF oscillators use resonant circuits that consist of a capacitor and coil (inductance) connected in series or parallel, as shown in Fig. 2.1. Such resonant circuits are used to tune the RF amplifier/oscillator network. At the resonant frequency, the inductive and capacitive reactances are equal, and the circuit acts as a high impedance (in a parallel circuit) or a low impedance (in a series circuit). In either case, any combination of capacitance and inductance has some resonant frequency.

Either (or both) the capacitance or inductance can be variable to permit tuning of the resonant circuit over a given frequency range. When the inductance is variable, tuning is often performed by a metal (powdered iron) slug inside the coil. The slug is screwdriver-adjusted to change the inductance (and thus the inductive reactance), as required.

Resonance and Impedance

Series (zero impedance) C —|— L

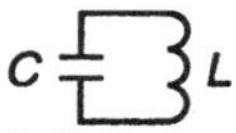

Parallel (infinite impedance) C ‖ L

$$F\,(\text{MHz}) = \frac{0.159}{\sqrt{L\,(\mu H) \times C\,(\mu F)}}$$

$$L\,(\mu H) = \frac{2.54 \times 10^4}{F\,(\text{kHz})^2 \times C\,(\mu F)}$$

$$C\,(\mu F) = \frac{2.54 \times 10^4}{F\,(\text{kHz})^2 \times L\,(\mu H)}$$

(a)

Capacitive Reactance

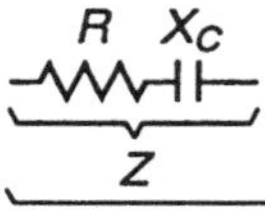

Series: R, X_C

$$Z = \sqrt{R^2 + X_C^2} \qquad Q = \frac{X_C}{R} \qquad C = \frac{1}{6.28 F X_C}$$

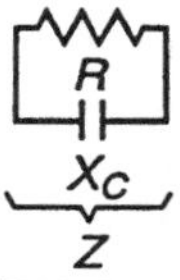

$$F = \frac{1}{6.28 C X_C}$$

Parallel: R, X_C

$$Z = \frac{R X_C}{\sqrt{R^2 + X_C^2}} \qquad Q = \frac{R}{X_C} \qquad X_C = \frac{159}{F\,(\text{kHz}) \times C\,(\mu F)}$$

(b)

Inductive Reactance

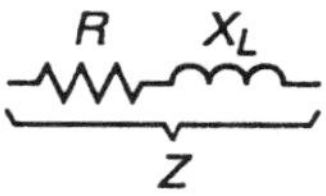

Series: R, X_L

$$Z = \sqrt{R^2 + X_L^2} \qquad Q = \frac{X_L}{R} \qquad L = \frac{X_L}{6.28 F}$$

$$F = \frac{X_L}{6.28 L}$$

Parallel: R, X_L

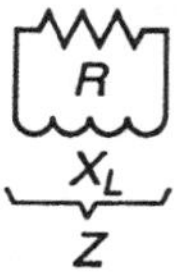

$$Z = \frac{R X_L}{\sqrt{R^2 + X_L^2}} \qquad Q = \frac{R}{X_L} \qquad X_L = 6.28 \times F\,(\text{MHz}) \times L\,(\mu F)$$

(c)

Figure 2.1 Resonant-circuit equations.

Typical RF circuits might include two resonant circuits in the form of a transformer (RF or IF transformers, for example). Again, either capacitance or inductance can be variable.

2.2 Resonant Frequency versus Q or Selectivity

All resonant circuits have a resonant frequency and a Q factor. The circuit Q depends on the ratio of reactance to resistance. If a resonant circuit has pure reactance, the Q is high (actually, infinite). However, this is not practical because any coil has some resistance, as do the leads of a capacitor. And, of course, a coil has resistance.

The resonant-circuit Q depends on the individual Q factors of inductance and capacitance used in the circuit. For example, if both the inductance and capacitance have high Q (very low resistance in relation to reactance), the circuit has high Q, provided that a minimum of resistance is produced when the inductance and capacitance are connected to form a resonant circuit.

Figure 2.1 has equations that show the relationships among capacitance, inductance, reactance, and frequencies, as these factors relate to resonant circuits. Notice the two sets of equations. One set includes reactance (inductive and capacitive); the other set omits reactance. The reason for the two sets of equations is that some design approaches require the reactance to be calculated for resonant networks.

From a practical standpoint, a resonant circuit with a high Q produces a sharp resonance curve (narrow bandwidth), whereas a low Q produces a broad resonance curve (wide bandwidth). For example, a high-Q resonant circuit provides good harmonic rejection and efficiency in comparison with a low-Q circuit, all other factors being equal.

The *selectivity* of a resonant circuit is related directly to Q. A very high Q (or high selectivity) is not always desired. Sometimes it is necessary to add resistance to a resonant circuit to broaden the response (increase the bandwidth and decrease selectivity). The resistances across the coils in Fig. 1.7 are examples.

Typically, resonant-circuit Q is measured at the point on either side of the resonant frequency where the signal amplitude is down 0.707 of the peak resonant value, as shown in Fig. 2.2. Notice that Q must be increased for increases in resonant frequency if the same bandwidth is to be measured. For example, if the resonant frequency is 10 MHz with a bandwidth of 2 MHz, the required circuit Q is 5. If the resonant frequency is increased to 50 MHz, with the same 2-MHz bandwidth, the required Q is 25. Circuit Q must be decreased for increases in bandwidth if the same resonant frequency is to be measured. For example, if the resonant frequency is 30 kHz, with a bandwidth of 2 kHz, the required circuit Q is 15. If the bandwidth is increased to 10 kHz, with the same 30-kHz resonant frequency, the required Q is 3.

2.3 Calculating Resonant Values

Assume that you want a circuit to resonate at 400 kHz with an inductance of 10 µH. What value of capacitor is necessary? Using the equations in Fig. 2.1,

$$C = \frac{2.54 \times 10^4}{400^2 \times 10} = 0.0158 \text{ µF}$$

Use the nearest standard value of 0.016 µF.

Assume that you want a circuit to resonate at 2.65 MHz with a capacitance of 360 pF. What value of inductance is necessary? Using the equations in Fig. 2.1,

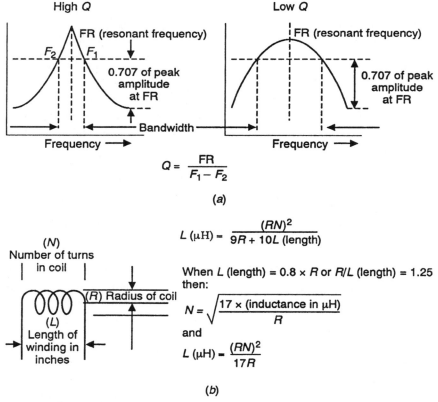

Figure 2.2 Q and inductance calculations for RF coils.

$$L = \frac{2.54 \times 10^4}{2650^2 \times (360 \times 10^{-6})} = 10 \ \mu H$$

Assume that you must find the resonant frequency of 0.002-μF capacitor and a 0.02-mH inductance. Using the equations in Fig. 2.1, first convert 0.02 mH to 20 μH, then:

$$F = \frac{0.159}{\sqrt{20 \ \mu H \times 0.002 \ \mu F}} = \frac{0.159}{\sqrt{0.04}} = \frac{0.159}{0.2}$$

$$= 0.795 \ \text{MHz, or } 795 \ \text{kHz}$$

Assume that an RF circuit must operate at 40 MHz with a bandwidth of 8 MHz. What circuit Q is required? Using the equations in Fig. 2.2,

$$F_R = 40 \quad F_1 - F_2 = 8 \quad Q = \frac{40}{8} = 5$$

2.4 Inductance of RF Coils

Figure 2.2 shows the equations necessary to calculate the self-inductance of a single-layer, air-core coil (the most common type of coil used in RF circuits). Notice that maximum inductance is obtained when the ratio of coil radius to coil length is 1.25 (when the length is 0.8 of the radius). RF coils for this ratio are the most efficient (maximum inductance for minimum physical size).

Assume that you must design an RF coil with 0.5-μH inductance on a 0.25-inch radius (air core, single layer). Using the equations in Fig. 2.2, for maximum efficiency, the coil length must be 0.8R, or 0.2 in. Then,

$$N = \sqrt{\frac{17 \times 0.5}{0.25}} = \sqrt{34} = 5.8 \ \text{turns}$$

For practical purposes, use six turns and spread the turns slightly. The additional part of a turn increases inductance, but the spreading decreases inductance. After the coil is made, check the inductance with an inductance bridge (or as described later in this chapter).

2.5 Voltage-Variable Capacitors (VVCs)

Many RF circuits are tuned with VVCs. Notice that a VVC is sometimes called a *voltage-variable diode* because the device is constructed more like a diode than a capacitor. Figures 1.8 and 1.10 show how VVCs (D153) are used to tune circuits in RF tuners.

The capacitance of a VVC is controlled by external voltage and is varied as the external voltage is varied (such as the error voltage applied to D153 in Fig. 1.10). If a VVC is used in an RF circuit, it is possible to vary the circuit capacitance and thus vary the resonant frequency (or tune the circuit) with a variable voltage. From a simplified-design standpoint, the main concern with VVCs in resonant circuits is the tuning range of the circuit. All other factors being equal, the tuning range depends on the capacitance range of the VVC.

2.6 Basic RF-Circuit Design Approaches

There are many approaches for the design of RF circuits. Two of the classic approaches are: *y-parameters* and *large-signal characteristics*. With either approach, it is difficult (at best) to provide simple, step-by-step procedures to meet all RF circuit conditions. In the practical world, there are several reasons why any step-by-step approach results in considerable trial and error. Here are some typical examples.

First, not all required design characteristics are always available in datasheet form. For example, input and output admittances for transistors might be given at some low frequency, but not at the desired operating frequency.

Often, manufacturers do not agree on terminology. A good example of this is in y-parameters for transistors, for which one manufacturer uses letter subscripts and another uses number subscripts (y_{fs} or y_{21}). Of course, this type of variation can be eliminated by conversion, but the process can be tedious.

In some cases, manufacturers give the required information on datasheets, but not in the required form. For example, some manufacturers might list the input capacitance of a transistor in farads, rather than listing the input in mhos. (The input admittance is then found when the input capacitance is multiplied by $6.28F$, where F is the frequency of interest. This conversion is based on the assumption that the input admittance is primarily capacitive and thus depends on frequency. The assumption is not always true for the frequency of interest. It might be necessary to use complex admittance-measuring equipment to make actual tests of the transistor.)

The input and output tuning circuit of an RF device (amplifier, multiplier, etc.) must perform three functions. First, the circuit elements (capacitors and coils) must tune the amplifier to the desired frequency. Second, the circuit elements must match the input and output impedances of the transistor to the impedances of the source and load (to minimize signal loss). Third, as in the case with any amplifier, some feedback always occurs between the output and input. If the admittance factors are just right, the feedback might be of sufficient amplitude and proper phase to cause oscillation in the circuit. The circuit is considered *unstable* when this occurs.

Circuit instability in any form is always undesirable and can be corrected by negative feedback (called *neutralization*) or by changes in the input-output tuning networks. Generally, the changes involve introducing some slight mismatch to improve stability. Although the neutralization and tuning circuits are relatively simple, the equations for determining stability (or instability) and impedance matching are long and complex (best solved by computer-aided design methods).

In an effort to cut through the maze of information and complex equations, simplified, practical, noncomputer-oriented RF-circuit design is covered. Armed with this information, you should be able to interpret datasheets or test information, and use the data to design tuning networks that provide stable RF-circuit operation at the frequency of interest. With each step, the various alternative procedures are covered and types of information available. Specific design examples are used to summarize the information. On the assumption that you might not be familiar with two-port networks, this section starts with a summary of the *y*-parameter system.

2.7 The *y*-parameter System

Impedance (Z) is a combination of resistance (R, the real part) and reactance (X, the imaginary part). Admittance (y) is the reciprocal of impedance, and is composed of conductance (g, the real part) and susceptance (jb, the imaginary part). Thus, g is the reciprocal of R, and jb is the reciprocal of X.

To find g, divide R into 1; to find R, divide g in 1. Z is expressed in ohms. y, being a reciprocal, is expressed in mhos or millimhos (mmhos). For example, an impedance of Z ohms equals 20 mmhos (1/50 = 0.02; 0.02 mho = 20 mmhos).

A *y*-parameter can be used as an expression for input admittance in the form:

$$y_i = g_i + jb_i$$

where g_i = real (conductive) part of input admittance
jb_i = imaginary (susceptive) part of input admittance
y_i = input admittance (the reciprocal of Z_i)

The term $y_i = g_i + jb_i$ expresses the *y*-parameter in *rectangular* form. Some transistor manufacturers describe the *y*-parameter in *polar* form. For example, they give the magnitude of the input as y_i and the angle of the input admittance as y_i. Quite often, manufacturers mix the two systems of *vector algebra* on datasheets.

2.7.1 Conversion of vector-algebra forms

In case you are not familiar with the basics of vector algebra, the following notes summarize the steps necessary to convert y-parameters expressed in vector-algebra terms. Of course, the actual conversion process requires the use of appropriate tables (sine, cosine, tangent, etc.).

To convert from rectangular to polar form:

1. Find the magnitude from the square root of the sum of the squares of the components:

$$\text{Polar magnitude} = \sqrt{g^2 + jb^2}$$

2. Find the angle from the ratio of the component values:

$$\text{Polar angle} = \arctan \frac{jb}{g}$$

The polar angle is leading if the jb term is positive and is lagging if the jb term is negative. For example, assume that y_{fs} is given as $g_{fs} = 30$ and $jb_{fs} = 70$.

$$|y_{fs}| \text{ polar magnitude} = \sqrt{30^2 + 70^2} = 76$$

$$y_{fs} \text{ polar angle} = \arctan \frac{70}{30} = 67°$$

Converting from polar to rectangular form:

1. Find the real (conductive, g) part when polar magnitude is multiplied by the cosine of the polar angle.
2. Find the imaginary (susceptance, jb) when polar magnitude is multiplied by the sine of the polar angle.

If the angle is positive, the jb component is also positive. When the angle is negative, the jb component is also negative.

For example, assume that the y_{fs} is given as $y_{fs} = |20|$ and $y_{fs} = -33°$. This is converted to rectangular form by:

$$20 \times \cos 33° = g_{fs} = 16.8$$

$$20 \times \sin 33° = jb_{fs} = 11$$

2.8 The Four Basic y-Parameters

Figure 2.3 shows the y-equivalent circuit for a field-effect transistor (FET). A similar circuit can be drawn for a bipolar transistor.

Notice that y-parameters can be expressed with number or letter subscripts. The number subscripts are universal and can apply to bipolar transistors, FETs, and IC RF circuits (such as described in

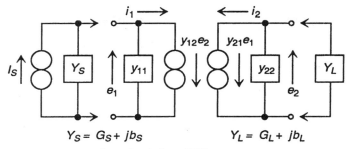

Figure 2.3 y-equivalent circuit for a FET.

Chapters 3 through 14). The letter subscripts are most popular on FET datasheets.

The following notes can be used to standardize y-parameter nomenclature. Notice that the letter s in the letter subscript refers to common-source operation of a FET circuit and is equivalent to a common-emitter bipolar circuit.

y_{11} is *input admittance* and can be expressed as y_{is}.

y_{12} is *reverse transadmittance* and can be expressed as y_{rs}.

y_{21} is *forward transadmittance* and can be expressed as y_{fs}.

y_{22} is *output admittance* and can be expressed as y_{os}.

2.8.1 Input admittance

When the load admittance (Y_L in Fig. 2.3) is infinite (the load is shorted), input admittance is expressed as:

$$y_{11} = g_{11} + jb_{11} = (di_1/de_1) \text{ (with } e_2 = 0\text{)}$$

This means that y_{11} is equal to the difference in current i_1, divided by the difference in voltage e_1, with voltage e_2 at 0. The voltages and currents involved are shown in Fig. 2.3.

Some datasheets do not show y_{11} at any frequency but give *input capacitance* instead. If it is assumed that the input admittance is entirely (or mostly) capacitive, the input impedance can be found when the input capacitance is multiplied by 6.28F (F = frequency in Hz) and the reciprocal is taken. Because admittance is the reciprocal of impedance, admittance can be found simply by multiplying the input capacitance by 6.28F (where admittance is capacitive). For example, if the frequency is 100 MHz and the input capacitance is 9 pF, the input admittance is:

$$6.28 \times (100 \times 10^6) \times (9 \times 10^{-12}) = \text{about 5.6 mmhos.}$$

This assumption is accurate only if the real part of y_{11} (or g_{11}) is negligible. Such an assumption is reasonable for most FETs, but not necessarily for all bipolar transistors. The real part of bipolar transistor input admittance can be quite large in relation to the imaginary jb_{11} part.

2.8.2 Forward transadmittance

When the load admittance (Y_L in Fig. 2.3) is infinite (the load is shorted), forward transadmittance is expressed as:

$$y_{21} = g_{21} + jb_{21} = (di_2/de_1) \qquad \text{(with } e_2 = 0\text{)}$$

This means that y_{21} is equal to the difference in output current i_2 divided by the difference in input voltage e_1, with e_2 at 0. In other words, y_{11} represents the difference in output current for a difference in input voltage.

Bipolar transistor datasheets often do not list any value for y_{21}. Instead, forward transadmittance is shown by a *hybrid system* of notation using h_{fe} or h_{21} (which means hybrid forward transadmittance with a common emitter). Regardless of what system is used, it is essential that the values for forward transadmittance be considered at the frequency of interest when designing RF networks.

2.8.3 Output admittance

When the input admittance (Y_S in Fig. 2.3) is infinite (the source or input is shorted), output admittance is expressed as:

$$y_{22} = g_{22} + jb_{22} = (di_2/de_2) \qquad \text{(with } e_1 = 0\text{)}$$

2.8.4 Reverse transadmittance

When the input admittance (Y_S in Fig. 2.3) is infinite (the source or input is shorted), reverse transadmittance is expressed as:

$$y_{12} = g_{12} + jb_{12} = (di_1/de_2) \qquad \text{(with } e_1 = 0\text{)}$$

2.9 Measuring y-Parameters

The main concern when measuring y-parameters is that the measurements are made under conditions simulating those of the final circuit. For example, if supply voltages, bias voltages, and operating frequency are not identical (or close) to the final circuit, the tests can be misleading. The most important y-parameters in RF-circuit design are input and output admittances. Although the datasheets for transistors to be used in RF circuits usually contain such admittance data, it can be helpful to know how this information is obtained.

Two basic methods are used to measure the y-parameters of transistors. One method involves *direct measurement* of the parameter (such as measuring changes in outputs for corresponding changes in input). The other method uses *tuning substitution* (where the transistor is tuned for maximum transfer of power, and the admittances of the tuning circuits are measured). Both methods are summarized in the following paragraphs.

2.9.1 Direct measurement of y_{fs} (y_{11})

Figure 2.4 shows a typical test circuit for direct measurement of y_{fs} (which might be listed as y_{21}, g_m, or even g_{fs}). Although an FET is shown, the same circuit can apply to any single-input device, such as a bipolar transistor or even the input of an RF IC.

The value of R_L must be such that the drop produced by the FET current drain is negligible, and the operating-voltage point (V_{DS}, in the case of an FET) is correct for a given power-supply voltage (V_{DD}) and operating current (I_D). For example, if I_D is 10 mA, V_{DD} is 20 V, and V_{DS} is 15 V, R_L must drop 5 V at 10 mA. Thus, the value of R_L is 5 V/0.01 A = 500 Ω.

During testing, the signal source is adjusted to the frequency of interest. The amplitude of the signal source (V_{in}) is set to some convenient number, such as 1 V or 100 mV. The value of y_{fs} (y_{11}) is calculated from the equation in Fig. 2.4, and is expressed in mhos (or millimhos and μmhos, in practical circuits). As an example, assume that the value of R_L is 1000 Ω, V_{in} is 1 V, and V_{out} is 8 V. The value of y_{fs} is:

$$8/(1 \times 1000) = 0.008 \text{ mho} = 8 \text{ mmho} = 8000 \text{ μmho}$$

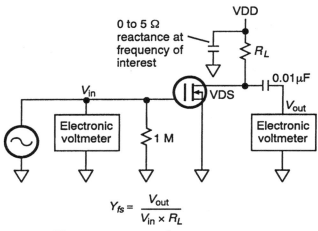

Figure 2.4 Direct measurement of y_{fs}.

2.9.2 Direct measurement of y_{os} (y_{22})

Figure 2.5 shows a typical test circuit for direct measurement of y_{os} (which might be listed as y_{22}, g_{os}, g_{22}, or even r_d, where $r_d = 1/y_{os}$). Some datasheets give y_{os} as a complex number with both the real (g_{os}) and imaginary (b_{os}) values shown by means of curves.

The value of R_S must be such as to cause a negligible drop (so that V_{DS} can be maintained at the desired level, with given V_{DD} and I_D). During testing, the signal source is adjusted to the frequency of interest. Both V_{out} and V_{DS} are measured, and the value of y_{os} is calculated from the equation in Fig. 2.5.

2.9.3 Direct measurement of y_{is} (y_{11})

Although y_{is} is not generally crucial, it is necessary to know the values of y_{is} to calculate impedance-matching networks. If it is necessary to establish the imaginary part (b_{is}), use an admittance meter or R_X meter.

2.9.4 Direct measurement of y_{rs} (y_{12})

Again, although y_{rs} is generally not crucial, it is necessary to know the values of y_{rs} to calculate impedance-matching networks. Although the real part of g_{rs} remains at zero for all conditions and at all frequencies, the imaginary part (b_{rs}) does vary with voltage, cur-

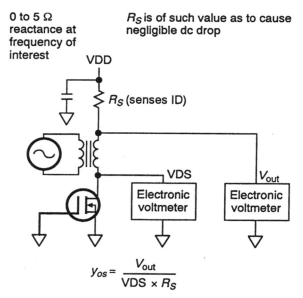

Figure 2.5 Direct measurement of y_{os}.

rent, and frequency. (The reverse susceptance varies and, under certain conditions, can produce undesired feedback from output to input.) This condition must be accounted for when designing RF amplifiers to prevent feedback from causing oscillation. If it is necessary to establish the imaginary part (b_{rs}), use an admittance meter or R_x meter.

2.9.5 Tuning-substitution measurement of y-parameters

Figure 2.6 is a typical test circuit for the measurement of y-parameters using the tuning-substitution method. Although a bipolar transistor is shown, the circuit can be adapted for use with FETs.

During testing, the transistor is placed in a test circuit with variable components to provide wide tuning capabilities. This is necessary to ensure correct matching at various power levels. The circuit

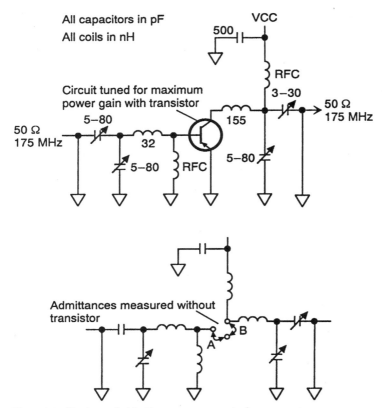

Figure 2.6 Tuning-substitution measurement of y-parameters.

is tuned for *maximum power gain* at each power level for which admittance information is desired. After the test amplifier is tuned for maximum power gain, the dc power, signal source, circuit load, and test transistors are disconnected from the circuit. For total circuit impedance to remain the same, the signal-source and output-load circuit connections are terminated at the characteristic impedances (typically 50 ohms).

After the substitutions are complete, complex admittances are measured at the base- and collector-circuit connections of the test transistor (points A and B, respectively, Fig. 2.6), using a precision admittance meter. The transistor input and output admittances are the *conjugates* of the base-circuit connection and the collector-circuit connection admittances, respectively. For example, if the base-circuit connection (point A) admittance is $8 + j3$, the input admittance of the transistor is $8 - j3$.

With the large-signal approach to RF-circuit design, the networks are calculated on the basis of input/output resistance and capacitance, instead of admittance (although admittances are often used to determine stability before going into design). An example of large-signal design is listed in Section 2.15. With the large-signal approach, the admittances measured in the test circuit of Fig. 2.6 are converted to resistance and capacitance.

Admittances are expressed in mhos (or mmhos). Resistance (in ohms) is found by dividing the real part of the admittance into 1. Capacitance (in farads) is found by dividing the imaginary part of the admittance into 1 (to find capacitive reactance in X_c). Then, the reactance is used in the equation $C = 1/(6.28\,F\,X_c)$, where C is the capacitance and F is the frequency of interest.

2.10 Using the Smith Chart

The characteristics of transistors and ICs used in RF work are often expressed in the form of Smith charts. It is also possible to design matching networks for RF circuits directly on Smith charts. The following paragraphs summarize the use of Smith charts as they apply to the ICs described in this book.

2.10.1 Smith chart construction

Figure 2.7 shows how a Smith chart is constructed. As shown, it has two sets of circles. One set comprises circles of *constant resistance* (Fig. 2.7A), and the other set has circles of constant *reactance* (Fig. 2.7B). The values of these circuits are normalized to the characteristic impedance of the system by dividing the actual value of resistance or

Practical Considerations for RF 39

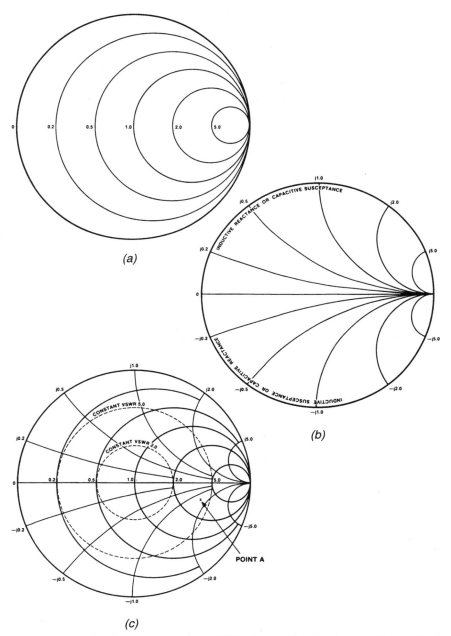

Figure 2.7 Smith chart construction (*GEC Plessey Semiconductors*, Professional Products IC Handbook, *p. 4-57*).

reactance by the characteristic impedance. For example, in a 50-Ω system, a resistance of 100 Ω is normalized to a value of 2.0.

A complete chart (such as shown in Fig. 2.7C) is formed when the circles of Figs. 2.7A and 2.7B are combined. Any normalized impedance has a unique position on the chart; the variation of this impedance with frequency or other parameters can be plotted directly. For example, an impedance of $150 - j75$ Ω can be represented by a normalized impedance (in a 50-Ω system) of $3 - j1.5$. This point is plotted in Fig. 2.7C as point A.

Typically, Smith charts also include circles of constant VSWR (voltage standing wave ratio), as shown in Fig. 2.7C (VSWR is covered further in this chapter). The main purpose of measuring and plotting VSWR is to show the degree of mismatch in a system. The VSWR is the ratio of the device impedance to the characteristic impedance, and is expressed as a ratio greater than 1. For example, a 25-Ω device (transistor or IC) in a 50-Ω system has a VSWR of 2:1.

2.10.2 Transforming impedances and admittances

As is covered further in this chapter, it is often easier to change a series RC network to its equivalent parallel network for calculation purposes. This is because, as a parallel network of admittances, a shunt admittance can be added directly (instead of with tedious calculations required for series networks). It is also easier to apply parallel networks. In general, it is easier to deal with admittances for shunt components, and reactances for series components. Either way, the Smith chart provides a simple graphic method for the transformation.

Any impedance can be represented (at a fixed frequency) by a shunt conductance and susceptance on the chart. By transferring that point on the chart to a point at the same diameter, but 180° away, the transformation is made automatically. This is shown in Fig. 2.8, where points A and B are the series and parallel equivalents of the same impedance.

2.10.3 Adding impedances and admittances

Figures 2.9 and 2.10 show how admittances and impedances can be added using a Smith chart. Assume that a series inductance is to be added to an admittance (for example, a parallel RC network). First, transform the admittance into a series impedance using the method shown in Fig. 2.8. The series inductance can then be added, as shown in Figs. 2.9 and 2.10.

Point A (on Fig. 2.10) is the starting admittance consisting of a shunt capacitance and resistance. The equivalent capacitive impedance is shown at point B. The addition of a series inductor moves the imped-

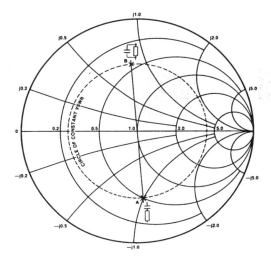

Figure 2.8 Series-reactance to parallel-admittance conversion (*GEC Plessey Semiconductors, Professional Products IC Handbook, p. 4-58*).

ance to point C. The value of this inductor is defined by the length of arc BC. In Fig. 2.10, the BC arc is from $-j0.5$ to $j0.43$ or a total of $j0.93$. (This reactance must, of course, be denormalized before it can be used in a practical circuit.)

Point C represents an inductive impedance, which is equivalent to the admittance shown at point D. The addition of the shunt reactance moves the input admittance to the center of the chart, and has a value of $-j2.0$. Point D should be chosen so that it lies on the unity impedance/conductance circle. This provides a focus of points for point C.

Keep two facts in mind when using Smith charts. First, the calculation is for one frequency only. If the frequency is changed, the reactance changes, making both the calculations and graphic representation subject to change. Second, Smith charts cannot provide the accuracy of a computer printout (or even that of a simple calculator). However, the chart method is simple and provides sufficient accuracy for first-trial values of RF networks. (If you prefer calculations, read the rest of this chapter!)

2.11 RF-Amplifier Stability

Many methods are used to determine the potential stability (or instability) of transistors in RF-amplifier circuits. One classic method, the Stern k factor, involves equations requiring y-parameter information.

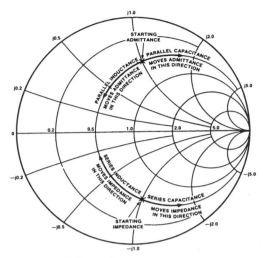

Figure 2.9 Effects of series and shunt reactance (*GEC Plessey Semiconductors,* Professional Products IC Handbook, *p. 4-58*).

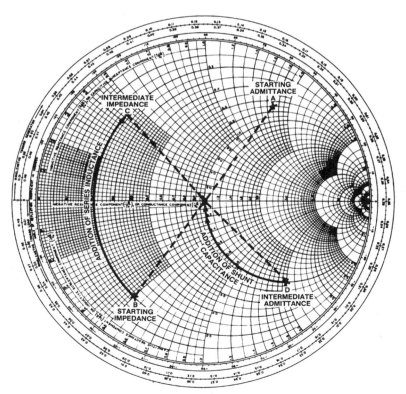

Figure 2.10 Matching design using the Smith chart (*GEC Plessey Semiconductors,* Professional Products IC Handbook, *p. 4-59*).

As mentioned, y-parameters can be taken from datasheets or by actual measurement at the frequency of interest.

The Stern k factor is calculated from:

$$k = \frac{2\,(g_{11} + G_S)(g_{22} + G_L)}{y_{12}y_{21} + R_e(Y_{12}y_{21})}$$

where G_S and G_L are source and load conductance, respectively (G_S = 1/*source resistance*; G_L = 1/*load resistance*).

If k is greater than 1, the amplifier circuit will be stable. If k is less than 1, the circuit is potentially unstable. In practical design, a Stern k factor of 3 or 4 is used, rather than 1, to provide a safety margin. This accommodates parameter and component variations (particularly with regard to bandpass response).

It is obvious that the calculations for Stern k are complex and best solved by computer (or a really first-rate calculator operator). Fortunately, some manufacturers provide alternative solutions to the stability and load-matching problems, usually in the form of datasheet graphs.

2.12 RF-Amplifier Stability Solutions

The two most common solutions to the problem of unstable RF-amplifier circuits are *neutralization* and *mismatch*. The following paragraphs summarize the solutions.

2.12.1 Neutralization

When an RF amplifier is neutralized, part of the output is fed back (after the output is shifted in phase) to the input, thus canceling any oscillation. With neutralization, an RF amplifier can be matched perfectly to the source and load. Neutralization is sometimes known as a *conjugate match*. In a perfect conjugate match, the transistor input and source, as well as the transistor output and load, are matched with regard to resistance, and all reactance is tuned out. The disadvantage to neutralization is that it requires extra components and can create a problem when the frequency is changed. (It might be necessary to readjust the neutralization circuit at each frequency change.)

2.12.2 Mismatch

The mismatch solution to amplifier instability involves introducing a specific amount of mismatch into either the source or load tuning networks so that any undesired feedback is not sufficient to produce instability or oscillation. The mismatch solution, sometimes called the *Stern solution* (because the Stern k factor is involved), requires no extra components. The disadvantage to mismatch is that it does reduce gain.

2.12.3 Mismatch versus neutralization

Figure 2.11 shows a comparison of the mismatch and neutralized solutions. The higher-gain curve represents neutralized operation (also called *unilateralized gain* in some literature). The lower-gain curve represents the power gain when the Stern k factor is 3.

Assume that the frequency of interest is 100 MHz. The datasheet graph of Fig. 2.11 shows that the circuit is potentially unstable if there is no neutralization or mismatch. If the RF circuit is matched directly to the load (perfect conjugate match) without regard to stability (or using neutralization to produce stability), the top curve applies and the power gain is about 38 dB. If the amplifier is matched to a load and source where the Stern k factor is 3 (resulting in a mismatch with the actual load and source), the lower curve applies, and the power gain is about 29 dB.

2.12.4 RF-amplifier gain calculations

Although there are many expressions used to show RF-amplifier gain, maximum available gain (MAG) and maximum usable gain (MUG) are most useful.

MAG is the gain in a conjugately matched, neutralized RF circuit and is expressed as:

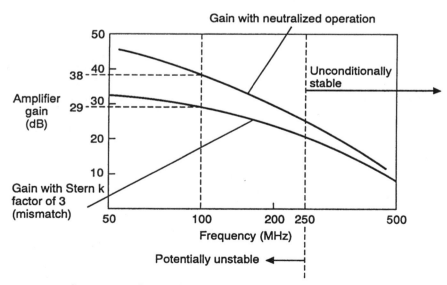

Figure 2.11 Comparison of neutralized and mismatched solutions for stability.

$$\text{MAG} = \frac{(y_{21})^2 R_{in} R_{out}}{4}$$

where R_{in} and R_{out} are the input and output, respectively, of the transistor.

An alternative MAG expression is:

$$\text{MAG} = \frac{(y_{21})^2}{4R_e(y_{11}) R_e(y_{22})}$$

where $R_e(y_{11})$ is the real part (g_{11}) of the input admittance, and $R_e(y_{22})$ is the real part (g_{22}) of the output admittance.

MUG is usually applied as the stable gain that can be realized in a practical (neutralized or unneutralized) RF amplifier. In a typical unneutralized circuit, MUG is expressed as:

$$\text{MUG} = \frac{0.4\,(y_{21})}{6.28\text{-F reverse transfer capacitance}}$$

MAG and MUG are often omitted on datasheets for bipolar transistors. Instead, gain is listed as h_{fe} at a given frequency. This is supplemented with graphs that show available power output at given frequencies with a given input.

2.13 RF Circuits with Neutralization

Several methods are used to neutralize RF amplifiers. The most common method is the *capacitance-bridge* technique shown in Fig. 2.12. Capacitance-bridge neutralization becomes more apparent when the circuit is redrawn, as shown in Fig. 2.12B. The condition for neutralization is that $I_F = I_N$ (the neutralization current I_N must be equal to the feedback current I_F in amplitude, but of opposite phase).

The equations normally used to find the value of the feedback neutralization capacitor are long and complex. However, for practical work, if C_1 is made quite large in relation to C_2 (at least four times), the value of C_N can be found by:

$$C_N = C_F\,(C_1/C_2)$$

where C_F is the reverse capacitance of the transistor.

In simple terms, C_N is about equal to the value of reverse capacitance times the ratio of C_1/C_2. For example, if reverse capacitance (sometimes listed as *collector-to-base capacitance*) is 7 pF, C_1 is 30 pF, and C_2 is 3 pF, the C_1/C_2 ratio is 10, and $C_N = (10)7$ pF = 70 pF.

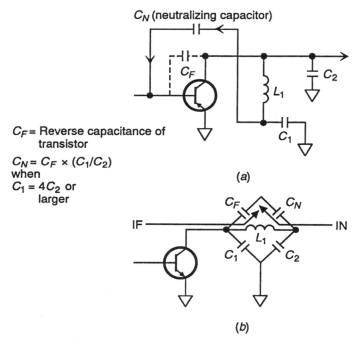

Figure 2.12 Capacitance-bridge neutralization for RF circuits.

2.14 RF Circuits without Neutralization (Mismatching)

A stable design with a potentially unstable transistor is possible without external feedback (neutralization) by proper choice of source and load values. This can be seen by inspection of the Stern k factor equation (Sec. 2.11). G_S and G_S can be made large enough (mismatched) to yield a stable circuit, without regard to potential instability. Using this approach, a circuit-stability factor (typically $k = 3$) is selected, and the Stern k factor equation is used to arrive at values of G_S and G_L that produce the desired k.

Of course, the actual G of the source and load cannot be changed. Instead, the input and output tuning circuits are designed as if the actual values are changed. This results in a mismatch and a reduction in power gain, but does produce the desired stability.

2.15 Large-Signal Design Approach

It is possible to design the tuning networks for RF circuits without using a full set of y-parameters or admittances. Instead, the networks are designed using the input/output capacitances and resistances of the transistor (the so-called *large-signal characteristics*). Often, the capacitance and resistance information is available on datasheets in the form of graphs. This is especially true for transistors used in RF power amplifiers and/or multipliers.

Assume that the circuit of Fig. 2.13 is to be used as an RF power amplifier with a 50-Ω antenna and a 28-V power supply. The circuit is to operate at 50 MHz and produce a 50-W output. Before getting into design calculations, here are some practical design considerations.

2.15.1 RF amplifiers versus multipliers

The same basic circuits in Fig. 2.13 can be used as RF amplifiers or frequency multipliers. However, in a multiplier circuit, the output must

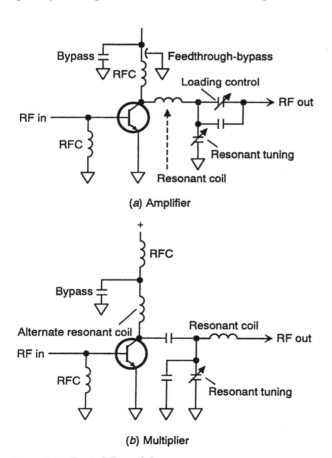

Figure 2.13 Basic RF amplifier.

be tuned to a multiple of the input. RF multipliers might not provide amplification. Usually, most of the amplification is supplied by the final amplifier stage, which is not operated as a multiplier. That is, the input and output of the final stage are at the same frequency. A typical RF transmitter has three stages: an oscillator to generate the basic signal frequency (Section 2.16), an intermediate stage that provides amplification and/or frequency multiplication, and a final stage for power amplification.

2.15.2 Tuning controls

The circuit of Fig. 2.13A has two tuning controls (variable capacitors, in this case) in the output network. The network in Fig. 2.13B has only one adjustment control. The circuit of Fig. 2.13A is typical for power amplifiers, where the output is tuned to the resonant frequency by one control and adjusted for proper impedance match by the other control (often called the *loading control*). In practice, both controls affect tuning and loading (impedance matching). The circuit in Fig. 2.13B is typical for multipliers or intermediate amplifiers where the main concern is tuning to the resonant frequency.

2.15.3 Parallel capacitors

Both networks make use of parallel capacitors. This parallel arrangement serves two purposes. First, there is a minimum fixed capacitance in case the variable capacitor is adjusted to minimum value. (In some cases, if a certain minimum capacitance is not included in the network, a severe mismatch can occur, possibly resulting in damage to the transistor.) The second purpose for a parallel capacitor is to reduce the required capacitance rating (and thus the physical size) of the variable capacitor.

2.15.4 Midrange capacitors

When designing networks (such as shown in Fig. 2.13), use a capacitor with a midrange capacitance equal to the desired capacitance. For example, if the desired capacitance is 25 pF (to produce resonance at the normal operating frequency), use a variable capacitor with a range of 1 to 50 pF. If such a capacitor is not readily available, use a fixed capacitor of 15 pF in parallel with a 15-pF variable capacitor. This provides a capacitance range of 16 to 30 pF, with a midrange of about 23 pF. Of course, the maximum capacitance range depends on the required tuning rate of the circuit. (A wide frequency range requires a wide capacitance range.)

2.15.5 Bias

Generally, RF power transistors remain cut off until a signal is applied. Therefore, the transistors are never conducting for more than 180° (half a cycle) of the 360° input signal cycle. In practice, the transistor conducts for about 140° of the input cycle, either on the positive or negative half, depending on the transistor type (NPN or PNP). No bias, as such, is required for this class of operation (class C).

2.15.6 Grounded emitter

The emitter is connected directly to ground. In those transistors where the emitter is connected to the case (typical in many RF power transistors), the case can be mounted on a metal surface (possibly part of the ground plane) that is connected to the ground side of the power supply. A direct connection between the emitter and ground is of particular importance in high-frequency operation. If the emitter is connected to ground through a resistance (or a long lead), an inductive or capacitive reactance can develop at high frequencies, resulting in undesired changes in the network.

2.15.7 RFC connections

The transistor base is connected to ground through an RF choke (RFC). This provides a dc return for the base, as well as RF signal isolation between the base and emitter or ground. The transistor collector is connected to the supply through an RFC and (in some cases) through the coil portion of the resonant network. The RFC provides dc return, but with RF signal isolation between the collector and power supply.

When the collector is connected to the supply through the resonant network, the coil must be capable of handling the full collector current. For this reason, final (power) amplifier networks should be chosen so that collector current does not pass through the coil. The circuit in Fig. 2.13A is, therefore, preferable to the circuit in Fig. 2.13B for power amplifiers. The circuit in Fig. 2.13B should be used where the current is low (such as in an intermediate amplifier).

2.15.8 RFC ratings

The ratings for RFCs are sometimes confusing. Some manufacturers list a full set of characteristics: inductance, dc resistance, ac resistance, Q, current capability, and normal frequency range. Other manufacturers list only one or two of these characteristics. Ac resistance and Q usually depend on frequency. A nominal frequency-range characteristic is a helpful, but usually not crucial, design parameter.

All other factors being the same, the dc resistance of an RFC should be at a minimum for any circuit carrying a large amount of current. For example, a large dc resistance in the collector of a final power amplifier can result in a large voltage drop between the supply and collector. Usually, the selection of a trial value for an RFC is based on a tradeoff between inductance and current capability. The minimum current capacity should be greater (by at least 10 percent) than the maximum anticipated direct current. The inductance depends on operating frequency. As a trial value, use an inductance that produces a reactance between 1000 and 3000 Ω at the operating frequency.

2.15.9 Bypass capacitors

The power-supply circuits of power amplifiers and multipliers must be bypassed, as shown in Fig. 2.13. The feed-through bypass capacitors are used at higher frequencies where the RF circuits are physically shielded from the supply and other circuits. A feed-through capacitor permits direct current to be applied through a shield, but prevents RF from passing outside the shield (RF is bypassed to the ground return). As a trial value, use a total bypass capacitance range of 0.001 to 0.1 μF.

2.15.10 Checking bypass capacitors and RFCs

If RF signals are present on the supply side of the line, the bypass capacitance and/or the RFC inductance are not adequate. A possible exception is where the RF signals are being picked up because of inadequate shielding. If the shielding is good and RF signals are present in the supply, increase the bypass capacitance value. As a second step, increase the RFC inductance. Be certain to check circuit performance with each increase in capacitance or inductance value. For example, too much bypass capacitance can cause undesired feedback and oscillation; too much RFC inductance can reduce amplifier output and efficiency.

2.15.11 Ferrite beads and microstrip circuits

Figure 2.14 shows some typical amplifier/multiplier circuits using ferrite beads and microstrip techniques. The ferrite beads provide additional RF signal isolation (on the ground side of base inductances of each amplifier/multiplier, in this case). Ferrite beads are often used where mechanical shielding is not practical.

The circuit shown in Fig. 2.14 provides 25 W of output power in the 450- to 512-MHz band, and is designed for 12.5-V operation. Figure 2.14B shows typical amplifier performance data. Figure 2.14C shows

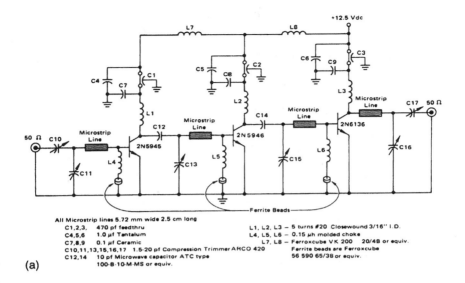

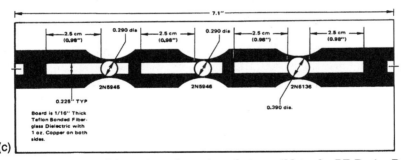

Figure 2.14 UHF amplifier using microstrip techniques (*Motorola*, RF Device Data, 1991, p. 7-51, 7-54).

the microstrip board layout. Those not familiar with microstrip techniques should read the many Motorola publications, such as AN548A and AN555. Notice that the circuits in Fig. 2.14 show all of the components included thus far in this section (RFCs, bypass capacitors, feed-through capacitors, variable capacitors, etc.).

2.15.12 Amplifier efficiency

A class C RF amplifier rarely has an efficiency over 65 to 70 percent. That is, the RF power output is 65 to 70 percent of the dc input power. To find the required dc input power, divide the desired RF power output by 0.6. For example, if the desired RF output is 50 W, the dc input power is 50/6 = 8.3, or about 83 W. Because the collector of an RF amplifier is at a dc potential about equal to the supply (slightly less because of a drop across the RFC and/or coil), divide the input power by the supply voltage to find the collector current.

2.15.13 Transistor characteristics

Probably the most important characteristic for RF amplifier transistors is that they must provide the necessary gain at the operating frequency. Of course, the transistors must also be capable of producing the required power output. Likewise, the input power to an amplifier must match the desired output and gain.

As an example, assume that a 50-W, 50-MHz transmitter is to be designed and that transistors with a power gain of 10 are available. Typically, a transistor oscillator produces less than a 1-W output. Therefore, an intermediate amplifier is required to deliver an output of 5 W to the final amplifier. The intermediate amplifier requires about 7-W dc input (50/7 = 7). Assuming a gain of 10 for the intermediate amplifier, an input of 0.5 W is required from the oscillator.

2.15.14 Intermediate-amplifier efficiency

When an intermediate amplifier is also used as a frequency multiplier, the efficiency drops. As a guideline, the efficiency of a second-harmonic amplifier (output at twice the input frequency) is about 42 percent, the third harmonic is 28 percent, the fourth harmonic is 21 percent, and the fifth harmonic is 18 percent. As an example, if an intermediate amplifier is to be operated at the second harmonic and to produce a 5-W RF output, the required dc input power is about 12 W (5/0.42 = 12).

2.15.15 Power gain with frequency multiplication

Another problem to be considered in frequency multiplication is that power gain (as listed on the datasheet) might not remain the same when the amplifier input and output are at different frequencies. Some datasheets specify power gain at the basic frequency, and then derate the power gain for second-harmonic operation. As a guideline, always use the minimum power-gain factor when calculating power input and output values.

2.15.16 Resonant-network design

The resonant network must be designed so that the network resonates at the desired frequency. That is, the inductive and capacitive reactance must be equal at the selected frequency. The network must also match the transistor output impedance to the load. Here are the main considerations.

A typical antenna-load impedance is about 50 Ω. The output impedance of a typical transistor is a few ohms (at radio frequencies). This mismatch will result in a loss of power to the load, and might cause damage to the transistor. The same problem exists when the output of one transistor (for example, an intermediate amplifier) does not match the input of another transistor (for example, the final amplifier).

Transistor impedance (both input and output) has both resistive and reactive components; thus, it varies with frequency. To design a resonant network for the output of a transistor, it is necessary to know the *output reactance* (usually capacitive), the *output resistance* at the operating frequency, and the *output power* (the so-called "large-signal parameters"). It is also necessary to know the input resistance and reactance (at a given frequency and power) when designing the resonant network of the stage feeding into a transistor.

2.15.17 Large-signal parameters

Figure 2.15 shows some typical large-signal parameters (in graph form) and the related equations. The reactance is found using the corresponding frequency and capacitance. For example, the output capacitance shown on the graph of Fig. 2.15 is about 12.5 pF at 80 MHz. This produces a capacitive reactance of about 159 Ω at 80 MHz. The reactance and resistance can then be combined to find impedance, as shown. Input and output impedances are generally listed on datasheets in parallel form. That is, it is assumed in the datasheets that the resistance is in parallel with the capacitance. However, some networks require that the impedance be calculated in series form, so it is necessary to convert using the equations in Fig. 2.15.

2.15.18 Large-signal resonant network calculations

Figure 2.16 shows five typical resonant networks, together with the calculations necessary to find the component values using large-signal data. Any of the resonant networks can be used as the tuning networks for RF amplifiers and/or multipliers. Notice that the network in Fig. 2.16A is similar to that in Fig. 2.13A, and Fig. 2.16C is similar to Fig. 2.13B (except for the power connections).

The resistor and capacitor shown in the box (labeled "Transistor to be matched") represent the complex output impedance of the transistor. When the network is to be used with a final amplifier, the resistor labeled R_L is the antenna impedance or other load. When the network is used with an intermediate amplifier, R_L represents the input impedance of the following transistor. It is, therefore, necessary to calculate the input impedance of the transistor being fed by the network, using the data and equations of Fig. 2.15.

The complex impedances are represented in series form in some cases and parallel form in others, depending on which form is most convenient for network calculation. The resultant network impedance, when terminated with a given load, must be equal to the conjugate of the impedance in the box. For example, assume that the transistor has a series output impedance of $7.33 - j3.87$. That is, the resistance (real part of impedance) is 3.87. For a maximum power transfer from the transistor to the load, the load impedance must be the conjugate of the output impedance, $7.33 + j3.87$.

R_S = Series resistance
X_S = Series reactance
$X_C = 1/(6.28FC)$

R_P = Parallel resistance
X_P = Parallel reactance
$X_L = 6.28FL$

Parallel output resistance = $\dfrac{\text{Collector voltage}^2}{2 \times \text{power output}}$

To convert X_S and R_S to parallel:

$$R_P = R_S\left[1 + \left(\dfrac{X_S}{R_S}\right)^2\right] \qquad X_P = \dfrac{R_P}{(X_S/R_S)}$$

To convert X_P and R_P to series:

$$R_S = \dfrac{R_P}{1 + (R_P/X_P)^2} \qquad X_S = R_S \dfrac{R_P}{X_P}$$

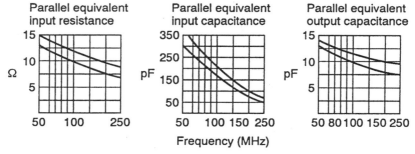

Figure 2.15 Typical large-signal graphs.

If the circuit is designed to operate into the typical 50-Ω load (antenna), the network must match the $(50 + j0)$-Ω load to the $7.33 - j3.87$ transistor value. In addition to matching, the network provides harmonic rejection to prevent transmission on more than one frequency (unless a harmonic is needed in a multiplier stage), low loss, and provisions for the adjustment of both loading and tuning.

2.15.19 Large-signal resonant network characteristics

Each large-signal network has advantages and disadvantages. The following is a summary of the five resonant networks shown in Fig. 2.16.

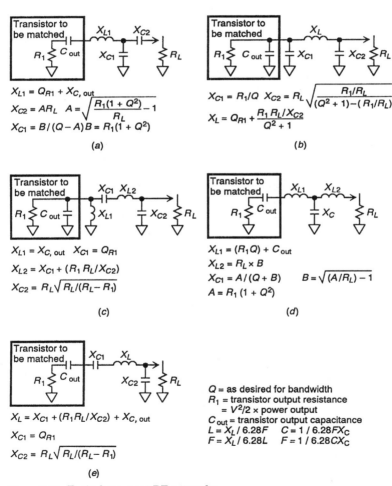

Figure 2.16 Typical resonant RF networks.

The network in Fig. 2.16A applies to most RF power amplifiers and is especially useful where the series real part of the transistor output impedance (R1) is less than 50 ohms. With a typical 50-Ω load, the required reactance for C1 rises to an impractical value when R1 is close to 50 ohms.

The network in Fig. 2.16B (often called a *pi network*) is best suited where the parallel resistance (R1) is high (near the value of R_L, typically 50 Ω). If the network in Fig. 2.16B is used with a low value of R1, the inductance of L1 must be very small, with C1 and C2 very large (beyond practical limits).

The networks in Fig. 2.16C and 2.16D produce practical values for C and L—especially where R1 is very low. The main limitation for the networks of Fig. 2.16C and 2.16D is that R1 must be substantially lower in d. These networks, or their many variations, are often used with intermediate stages where a low output impedance of one transistor is matched to the low input impedance of another transistor.

The network in Fig. 2.16E (often called a *tee network*) is best suited where R1 is much less (or much greater) than RL.

2.15.20 Practical large-signal resonant network

Assume that a network similar to that of Fig. 2.16A must match a transistor to a 50-Ω antenna with an operating frequency of 50 MHz. The required output is 50 W with a 28-V power supply. The transistor output capacitance C_{out} is 200 pF at 50 MHz (obtained from the datasheet or by testing).

With a power supply of 28 V and an output of 50 W, Fig. 2.15 shows that the parallel output resistance R1 is: $28^2/(2 \times 50) = 7.84$ Ω.

With an output capacitance of 200 pF and an operating frequency of 50 MHz, Fig. 2.15 shows the reactance of C_{out} as:

$$\frac{1}{6.28 \times (50 \times 10^6) \times (200 \times 10^{-12})} = 16 \; \Omega$$

The combination of these two values results in a parallel output impedance of $7.84 - j16$.

Usually, the datasheet lists the output capacitance in parallel form with R1. For the network in Fig. 2.16A, the values of R_1 and C_{out} must be converted to series form.

Using the equations in Fig. 2.15, the equivalent series output impedance is:

$$R_{series} = \frac{7.84}{1 + (7.84/16)^2} = 6.32 \; \Omega \; (R_1)$$

$$X_{series} = 6.32 \times \frac{7.84}{16} = 3.1 \; \Omega \; (C_{out})$$

The combination of these two values results in a series output impedance of $6.32 - j3.1$.

Using the equations in Fig. 2.16 and assuming a Q of 10 (for simplicity), the reactance values for the network are:

$$X_L = 10 \times (6.32) + 3.1 = 66.3 \; \Omega$$

$$A = \sqrt{\frac{6.32(1 + 10^2)}{50} - 1} = 3.4$$

$$X_{C2} = 3.3 \times 50 = 170 \; \Omega$$

$$B = 6.32(1 \times 10^2) = 638.32$$

$$X_{C1} = \frac{623.32}{10 - 3.4} = 96.7 \; \Omega$$

Using the equations in Fig. 2.16, the corresponding inductance and capacitance values are:

$$L_1 = \frac{66.3}{6.28 \times (50 \times 10^6)} = 0.21 \; \mu H$$

$$C_1 = \frac{1}{6.28 \times (50 \times 10^6) \times (96.7)} = 33 \; pF$$

$$C_2 = \frac{1}{6.28 \times (50 \times 10^6) \times (170)} = 19 \; pF$$

If C_1 and C_2 are variable, the values obtained should be the midrange values.

2.16 Discrete-Component RF-Oscillator Circuit Design

This section provides a review and summary of RF-oscillator design using discrete-component circuits. Chapter 6 describes IC RF oscillators used in wireless communications.

2.16.1 Basic RF-oscillator design considerations

The main concern in any oscillator design is that the transistor oscillates at the desired frequency and produces the desired voltage or power. Most discrete-component oscillator circuits operate with power outputs of less than 1 W, as do most wireless/RF ICs. Such devices can handle this power dissipation without heatsinks, so heatsinks are not covered extensively in this book. (For a thorough discussion of heatsinks, read *Lenk's RF Handbook*, McGraw-Hill, 1992.) However, temperature-related problems for both transistors and ICs are included in Section 2.17. PC-board layout considerations for transistors and ICs are included in the related chapters (3 through 14).

2.16.2 LC and crystal-controlled oscillators

Figure 2.17A and 2.17B shows two classic LC oscillators (the Hartley and Colpitts, respectively). LC oscillators are those that use inductances (coils) and capacitors as the frequency-determining components, and are still used in discrete-component RF circuits. Typically, the coils and capacitors are connected in series- and parallel-resonant circuits, and can be adjusted to the desired operating frequency. Either the coil or capacitor can be variable.

The main problem with a basic LC oscillator is that the frequency can drift (with changes in temperature, power-supply voltage, mechanical vibration, etc.). This can be overcome by crystal control, where a quartz crystal is used to set the operating frequency. Typically, the LC circuit is then adjusted to "trim" the oscillator output to an exact frequency.

Figure 2.18 shows a typical LC oscillator with crystal control. This circuit is one of the many variations of the Colpitts oscillator, where the output frequency is fixed and controlled by the crystal. The circuit can be tuned over a narrow RF range by a slug-tuned L1.

Figures 2.17C and 2.16D show two classic versions of the Pierce oscillator. Such circuits are very popular in RF work because of their simplicity and the minimum number of components. No LC circuits are required for frequency control. Instead, the frequency is set by the crystal alone.

2.16.3 RF-oscillator design considerations

Many factors must be considered in the design of RF oscillators. For example, the frequency-determining components must be temperature stable, and mechanical movement of the individual components must not be possible. The following paragraphs provide a summary of the design considerations for an oscillator (such as that shown in Fig. 2.18).

2.16.4 Frequency stability

Many factors affect oscillator frequency stability. For example, some optimum bias value and supply voltage usually produce maximum frequency stability over a given range of operating temperature. However, the one factor that can be controlled by the designer is percentage of feedback. (Note that this percentage refers to feedback versus output voltage.)

2.16.5 Feedback percentage

The lowest practical feedback level is about 10 percent. Rarely is more than 25 percent ever required, although some oscillators are operated at 40 percent. If the operating frequency is in the VHF or UHF regions

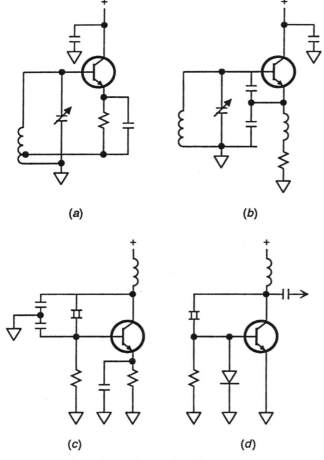

Figure 2.17 Classic LC and crystal oscillators.

Crystal frequency	RF power output at a percentage of dc power input			
	Fund	Harmonics		
		Second	Third	Fourth
Fundamental	30	15	10	5
Third overtone	25	15	10	5
Fifth overtone	20	12	7	3
Seventh overtone	20	12	7	3

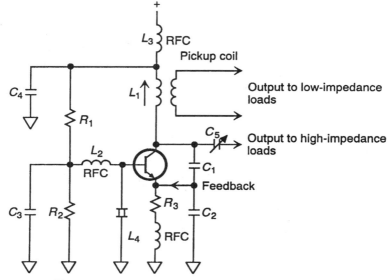

Figure 2.18 Typical LC oscillator with crystal control.

(see Fig. 1.1), the percentage of feedback must be increased over that of a comparable oscillator operating at low frequencies. Likewise, the percentage of feedback must be increased if the tuning circuits are made high C.

2.16.6 High C versus low C

The resonant frequency of RF oscillator circuits is usually set by the combination of L and C values. If the value of C is made quite large (with a corresponding lower value of L), the resonant circuit is said to be *high C*, and it usually results in sharper resonant tuning. A large value of L (with corresponding lower value of C) introduces more resistance into the resonant circuit, thus lowering the circuit Q to produce broader resonant tuning.

2.16.7 Crystal frequency

The resonant circuit (C1, C2, L1, and the transistor output capacitance in Fig. 2.18) should be at the same frequency as the crystal for maximum efficiency (maximum power output for a given supply voltage and current). If reduced efficiency is acceptable, the resonant circuit can be at a higher frequency (multiple) of the crystal frequency. However, the resonant circuit should not be operated at a frequency higher than the fourth harmonic of the crystal frequency.

2.16.8 Bias circuit

The bias-circuit components (R1, R2, and R3 in Fig. 2.18) are selected to produce a given current flow under no-signal conditions. The bias circuit is calculated on the basis of normal operating point—even though the circuit is never at the operating point. A feedback signal is always present, and the transistor is always in a state of transition. The collector current should be set to a value to produce the required output power. With the correct bias-feedback relationship, the output power of the oscillator is about 0.3 times the input power.

Typically, the voltage drop across L1 and L3 is very small, so the collector voltage equals the supply voltage. Thus, to find a correct value of current for a given power output and supply voltage, divide the desired output by 0.3 to find the required input power. Then divide the input power by the supply voltage to find the collector-current voltage.

2.16.9 Feedback signal

The amount of feedback is set by the ratio of C1 and C2 (Fig. 2.18). For example, if C1 and C2 are of the same value, the feedback signal is one half of the output signal. If C2 is made about three times the value of C1, the feedback signal is about 0.25 of the output signal voltage.

It might be necessary to change C_1 in relation to C_2 for a good bias-feedback relationship. For example, if C_2 is decreased, the feedback increases, and the oscillator operates nearer the class C region (where the transistor is cut off for about one half of each cycle). An increase in C_2, with C_1 fixed, decreases the feedback and makes the oscillator class A (where the transistor is never cut off). Remember that any change in C_2 (or C_1) also affects frequency. If the C_1 and C_2 values are changed, it will probably be necessary to change the value of L_1.

As a first trial value, the amount of feedback should be equal to, or greater than, cutoff. That is, the feedback voltage should be equal to or greater than the voltage necessary to cut off collector current flow. Under normal conditions, such a level of feedback should be sufficient to overcome the fixed bias (set by R_1 and R_2) and the variable bias set by R_3. As covered, feedback is generally within the limits of 10 and 40 percent, with the best stability in the 15 to 25 percent range.

2.16.10 Frequency

The frequency of the circuit in Fig. 2.18 is determined by the resonant frequency of L, C_1, and C_2 and by the crystal frequency. Notice that C1 and C2 are in series, so the total capacitance is found by the conventional series equation: $C = (C_1 \times C_2)/(C_1 + C_2)$. Also notice that the output capacitance of the transistor must be added to the value of C_1. At low frequencies, the output capacitance can be ignored because the value is usually quite low in relation to a typical value for C_1. At higher frequencies, the value of C_1 is lower, so the output capacitance becomes more important. (Transistor output capacitance can be considered as being in parallel with C1.)

The capacitance presented by the transistor output (collector-to-emitter) is composed of both output capacitance and reverse capacitance. However, reverse capacitance is usually small in relation to output capacitance, and can generally be ignored. When output capacitance is not available on datasheets, it is possible to calculate an approximate value of output admittance (the jb part represents susceptance, which is the reciprocal of reactance), and then find the corresponding capacitance.

As an example, to find the reactance presented by the collector-emitter terminals of the transistor at the datasheet frequency, first divide the jb part into 1. Then find the capacitance that produces such reactance at the datasheet frequency using the equation $C = 1/(6.28 \times F_{XC})$, where C is the output capacitance, F is the frequency, and XC is the capacitive reactance found as the reciprocal of the jb part of the output admittance. Of course, this method assumes that the jb reactance is capacitive and that the capacitance remains constant at all frequencies (at least the datasheet capacitance is the same for the design frequency). Neither of these conditions is always true.

In a practical circuit, it is generally easier to make L_1 variable, rather than C_1, because the tuning range of a crystal-controlled oscillator is quite small. Typically, the value of C_1 is about three times the value of C_1 (or the combined values of C1 and the transistor output capacitance, where applicable). With this ratio, the signal voltage (fed back from the emitter terminal) is about 0.25 of the total output signal voltage (or about 0.2 of the supply voltage), when the usual bias-feedback relationship is established.

2.16.11 L/C combinations for resonant circuits

Any number of L and C combinations can be used to produce the desired frequency. That is, the coil can be made very large or very small, with corresponding capacitor values. Often, practical limitations are placed on the resonant circuit (such as available variable-inductance values). In the absence of some specific limitations, and as a starting point for resonant-circuit values, the capacitance should be 2 pF per meter. For example, if the frequency is 30 MHz, the wavelength is 10 m, and the capacitance should be 20 pF. The wavelength in meters (m) is found by the equation: *wavelength* (meters) = 300/frequency (MHz).

At frequencies below about 1 to 5 MHz, the 2-pF-per-m guideline might result in very large coils to produce the corresponding inductances. If so, the 2 pF can be raised to 20 pF-per-meter. As an alternative method to find realistic values for the resonant circuit, use an inductive-reactance value (for L1) between 80 and 100 Ω at the operating frequency. This guideline is particularly useful at low frequencies (below 1 MHz).

2.16.12 Output circuit

The output to the following stage can be taken from L1 via a pickup coil (for low-impedance loads) or coupling capacitor (for high-impedance loads). Generally, the most convenient output scheme is to use a coupling capacitor (C5) and make the capacitor variable. This makes it possible to couple the oscillator to a variable load (a load that changes impedance with changes in frequency).

2.16.13 Crystal

The crystal must, of course, be resonant at the desired operating frequency (or a submultiple thereof, when the circuit is used as a multiplier). The efficiency (power output in relation to power input) of the oscillator is reduced when the oscillator is also used as a multiplier. This is shown in the table of Fig. 2.18, which also illustrates that efficiency is reduced when overtone crystals are used (instead of fundamental crystals). The crystal must be capable of withstanding the combined dc and signal voltages at the transistor input (base). As a guideline, the crystal should be capable of withstanding the full supply voltage—even though the crystal is never operated at this level.

2.16.14 Bypass and coupling capacitors

The values of bypass capacitors C3 and C4 should be such that the reactance is 5 Ω or less at the crystal operating frequency. A higher reactance (possibly up to 200 Ω) could be tolerated. However, because of the low output from most crystals, the lower reactance is preferred. The value of C5 should be about equal to the combined parallel output capacitance of the transistor and C1. Make this the midrange value of C5 (if C5 is variable).

2.16.15 RFCs

The values of L2, L3, and L4 (RF chokes) should be such that the reactance is between 1000 and 3000 Ω at the operating frequency. The minimum current capacity of the chokes should be greater (by at least 10 percent) than the maximum anticipated direct current. Note that a high reactance is desired at the operating frequency. However, at high frequencies, this can result in very large chokes that produce a large voltage drop (or are too large physically).

2.16.16 Crystal-oscillator design example

Assume that the circuit of Fig. 2.18 is to provide an output at 50 MHz. The circuit is to be tuned by L1, and a 30-V supply is available. The crystal is not damaged by 30 V and operates at 50 MHz with the desired accuracy. The transistor has an output capacitance of 3 pF and operates without damage at 30 V. The desired output power is 40 to 50 mW.

The collector is operated at 30 V (ignoring the small drop across L1 and L3). R_1, R_2, and R_3 should be chosen to provide a current that produces a 40- to 50-mW output with 30 V at the collector. A 45-mW output divided by 0.3 is 150 mW. Thus, the input power (and total dissipation) is 150 mW. Be certain that the transistor permits a 150-mW dissipation at the maximum anticipated temperature. Refer to Section 2.17 for thermal design considerations.

Assume that the transistor has a 330-mW maximum power dissipation at 25°C, a maximum temperature rating of 175°C, and a 2-mW/°C derating for temperatures above 25°C. If the transistor is operated at 100°C (75° above the 25°C level), the transistor must be derated at 150 mW (75 × 2 mW/°C), or 330 − 150 mW = 180 mW. Under these conditions, the 150-mW input power dissipation is safe.

With 30 V at the collector, and a desired 150-mW input power, the collector current must be 150 mW/30 V = 5 mA. Bias resistors R1, R2, and R3 should be selected to provide this 5-mA collector current using conventional methods. Keep in mind that, for a silicon NPN transistor shown in Fig. 2.18, the base voltage (set by R1 and R2) should be about 0.6 V higher than the emitter voltage (set by R3). For example, make R_1 equal to R_2 so that the base is set at 15 V. Then adjust R3 until the desired 5 mA of emitter-collector current is flowing. (Start out with about 2800 Ω for R3.)

With a 30-V supply, the output signal is about 24 V (30 × 0.8 = 24). Of course, this depends on the bias-feedback relationship. Also, the collector current does not remain at 5 mA when the circuit is oscillating because the transistor is always in a state of transition.

As a starting point, make C_2 three times the value of C_1 (plus the transistor output capacitance). With this ratio, the feedback is about 35 percent of the output or 6 V (24 × 0.25 = 6). Considering the amount of fixed and variable bias supplied by the bias network, a feedback of 6 V might be large. However, the 6-V value should be a good starting point.

For realistic values of L and C in the resonant circuit, let $C_1 = 2$ pF per meter, or 12 pF (300/50 = 6; 6 meters times 2 pF = 12 pF). With C_1 at 12 pF and the transistor output capacitance at 3 pF, the value of C_2 is 45 pF (12 + 3 = 15; 15 × 3 = 45).

The total capacitance across L1 is: (15 × 45)/(15 + 45) = 11.25 pF. With a value of 11.25 pF across L1, the value of L1 for resonance at 50 MHz is $(2.54 \times 10^4)/(50^2 \times 11.5) = 0.9$ μH. For convenience, L1 should be tunable from about 0.5 to 1.5 μH.

Remember that an incorrect bias-feedback relationship causes distortion of the oscillator waveform, low power, or both. The final test of the correct operating point is a good waveform at the operating frequency, together with frequency stability at the desired output power. Also, feedback is set by the relationship of C_1 and C_2. Any change in this relationship (to change the oscillator characteristics) requires a corresponding change in L_1. As a guideline, if no realistic combination of C_1, C_2, and L_1 produces the desired waveform and power output (but with the correct frequency), try a change in bias (controlled by R_1, R_2, and R_3).

The values of C3 and C4 should be: 1/(6.28 × 50 MHz × 5) = 630 pF. A slightly larger value (for example, 1000 pF) ensures that the reactance is less than 5 Ω (about 3 Ω) at the operating frequency.

The values of L2, L3, and L4 should be: 2000/(6.28 × 50 MHz) = 6.3 μH. Any value between about 3 and 9 μH should be good. The best test for the correct value of an RFC in an RF oscillator is to check for RF at the power-supply side of the line, with the oscillator operating.

As an example, check for RF at the point where L3 connects to the power supply (not at the L1 side). There should be no RF, nor should RF be greater than a few microvolts for a typical transistor oscillator (such as described in this example) at the supply side of L3. Finally, check for dc-voltage drop across the choke. The drop should be a fraction of 1 V (typically in the microvolt range).

2.17 Thermal Design Consideration

Three crucial parameters for transistors used in RF circuits are current gain, collector leakage, and power dissipation (in addition to output capacitance). These parameters change with temperature. To compound the problem, a change in parameters can also affect temperatures (for example, an increase in current gain or power dissipation results in a temperature increase). All of these problems can combine to produce *thermal runaway*.

Heat is generated when current passes through a transistor junction or the junctions on an IC. If all heat is not dissipated by the case (often an impossibility), the junction temperature rises. This, in turn, causes more current to flow through the junction—even though the voltage and circuit values remain the same. With more current, the junction temperature increases even further, with a corresponding increase in current flow. The transistor or IC burns out if the heat is not dissipated by some means.

When power dissipation is greater than about 1 W for a single transistor or IC, heatsinks (or special mounting provisions) are used to offset thermal runaway. For example, if a transistor or IC is used with a heatsink (or is mounted on a metal surface that acts as a heatsink), an increase in temperature (from any cause) can be dissipated into the surrounding air.

2.17.1 Effects of temperature

Collector leakage increases with temperature. As a guideline, collector leakage doubles with every 10°C increase in temperature for germanium devices, and doubles every 15°C for silicon. Also, always consider the possible effects of a different collector voltage when approximating transistor collector leakage at temperatures other than those on the datasheet.

Current gain increases with temperature. As a guideline, current gain doubles when the temperature is raised from 25 to 100°C for germanium and doubles when the temperature is raised from 25 to 175°C for silicon. If the datasheet does not specify a maximum operating temperature (or if there is no datasheet), do not exceed 100°C for germanium and an absolute maximum of 200°C for silicon (although 175°C is much safer).

If a heatsink is required, never apply power during service or design with the heatsinks removed or you will quickly learn the effects of temperature on power dissipation.

RF components (particularly power transistors) often have some form of *thermal resistance* specified to show the power-dissipation capabilities. Thermal resistance can be defined as the increase in temperature of the component junction (with respect to some reference) divided by the power dissipated, or degrees centigrade per watt (°C/W). In RF power transistors, thermal resistance is normally measured from the junction to the case, resulting in the term θ_{JC} (the lower-case Greek letter theta is used because engineers like Greek letters). When the case is bolted directly to the mounting surface with a built-in threaded bolt or stud, the terms θ_{MB} (thermal resistance to mounting base) or θ_{MF} (thermal resistance to mounting flange) are used.

Maximum power dissipation is specified in a variety of ways on datasheets. Some manufacturers provide *safe-operating-area* curves or graphs for temperature and/or power dissipation. Other manufacturers specify *maximum power dissipation,* in relation to a given ambient temperature (T_A) or a given case temperature. Still others specify a *maximum junction temperature* or *maximum case temperature.*

2.17.2 Calculating power dissipation for discrete transistors

The no-signal dc collector voltage and current can be used to calculate power dissipation when a transistor is operated under steady-state conditions (such as an RF amplifier or oscillator). Other calculations must be used for pulse operating conditions. In theory, other currents produce power dissipation (collector-base leakage current, emitter-base current, etc.). However, these can be ignored, and power dissipation (in watts) can be considered as the dc collector voltage times the collector current.

When the power dissipation is calculated, the *maximum power dissipation* must then be found. Under steady-state conditions, the maximum dissipation capability depends on:

1. The sum of the series thermal resistances from the transistor junction to ambient air.
2. The maximum junction temperature.
3. The ambient temperature.

Assume that it is desired to find the maximum power dissipation of a transistor under the following conditions: a maximum junction temperature of 175°C, a junction-to-case thermal resistance of 2°C/W, a heatsink with a thermal resistance of 3°C/W, and an ambient temperature of 25°C.

First, find the total junction-to-ambient thermal resistance: 2°C/W + 3°C/W = 5°C/W.

Next, find the maximum permitted power dissipation: (175°C − 25°C)/(5°C/W) = 30 W (maximum).

If the same transistor is used without a heatsink, but under the same conditions and with a TO-3 case (which is rated at 30°C/W), the maximum power can be calculated as follows:

First, find the total junction-to-ambient thermal resistance: 2°C/W + 30°C/W = 32°C/W.

Next, find the maximum permitted power dissipation: (175°C = 25°C)/(32°C/W) = 4.7 W (approximate).

Some transistor datasheets specify a *maximum case temperature*, rather than a maximum junction temperature. Assume that a maximum case temperature of 130°C is specified instead of a maximum junction temperature of 175°C. In that event, subtract the ambient temperature from the maximum permitted case temperature: 130°C − 25°C. Then divide the case temperature by the heatsink thermal resistance: 105°C/3°C/W = 35 W maximum power.

2.17.3 Calculating power dissipation for ICs

In the case of ICs, it is quite common to list only the maximum power dissipation for a given ambient temperature, and then show a *derating factor* in terms of maximum power decrease for a given increase in temperature. For example, the tuner IC covered in Chapter 3 has a maximum power dissipation of 1002 mW (1.02 W) at an ambient temperature of 70°C, with a derating factor of 12.5 mW/°C for ambient temperatures greater than 70°C. If the IC is operated at 100°C ambient, the maximum power is as follows: 100°C − 70°C = 30°C; 30 × 12.5 mW = 375 mW; 1002 mW − 375 mW = 627 mW. If the ambient is increased to 125°C, then the maximum power is: 125°C − 70°C = 55°C; 55 × 12.5 mW = 687.5 mW; 1002 mW − 687.5 mW = 314.5 mW.

From a practical design standpoint, an IC is a complete, predesigned, functioning circuit that cannot be altered in regard to power dissipation. If the supply voltages, input signals, output loads, and ambient temperature are at the recommended levels, the power dissipation will be within the IC's capabilities. With the possible exception of the data required to select heatsinks (as covered in Section 2.17.2), the designer need only follow the datasheet recommendations.

2.18 RF Test Equipment

It is assumed that you are familiar with basic test equipment, so such information is not repeated here. The test equipment used in RF work includes signal generators, scopes, meters, probes, frequency meters and counters, wattmeters, and dummy loads. (Never operate an RF circuit without a load connected at the output. This will almost certainly cause damage to the circuit, and possibly cause interference at the circuit frequency.) Two testing techniques are often used in RF design: *standing-wave ratio* measurement and *spectrum analysis*. Because the results of such tests are often listed on RF IC datasheets to show the relative merits of a particular IC, these classic measurements are included.

2.18.1 Standing-wave-ratio measurement

The standing-wave ratio of an antenna is actually a measure of matching or mismatching the antenna, transmission line (lead-in), and transmitter or RF output circuit. When the impedances of the antenna, line, and RF circuit are perfectly matched, all of the energy (or RF signal) is transferred to or from the antenna, and there is no loss. If a mismatch occurs (as is the case in any practical application), some of the energy is reflected back into the line. This energy cancels part of the RF signal.

If the voltage (or current) is measured along the line, there are voltage or current maximums (where the reflected signals are in phase with the outgoing signals) and voltage or current minimums (where the reflected RF signal is out of phase, partially canceling the outgoing signal). The maximums and minimums are called *standing waves*. The ratio can be related to either voltage or current. Because voltage is easier to measure, voltage is usually used (except in some RF lab work), resulting in the common term *voltage standing-wave ratio* (VSWR) found on RF IC datasheets.

The theoretical calculations for VSWR are shown in Fig. 2.19A. An SWR of 1 to 1, expressed as 1:1, means that there are no maximums or minimums (the voltage is constant at any point along the line) and that the match is perfect for circuit, line, and antenna. As a practical matter, if this 1:1 ratio should occur on one frequency, the ratio will not occur at any other frequency because impedance changes with frequency. It is not likely that all three elements (circuit, antenna, and line) will change impedance by exactly the same amount on all frequencies. Therefore, when checking VSWR, always check on all frequencies or channels, where practical. As an alternative, check SWR at the high, low, and middle channels (frequencies).

In the case of microwave RF signals being measured in the lab, a meter is physically moved along the line to measure maximum and minimum voltages. This is not practical for most wireless/RF equipment. It is far more convenient to measure forward or outgoing voltage, and reflected voltage, and then calculate the *reflection coefficient* (reflected/outgoing voltages). For example, using a 10-V forward and a 2-V reflected voltage, the reflection coefficient is 0.2.

Reflection coefficient is converted to SWR by dividing (1 + *reflection coefficient*) by (1 − *reflection coefficient*). For example, using the 0.2 reflection coefficient, the SWR is (1 + 0.2)/(1 − 0.2) = 1.2/0.8 = 1.5 SWR. This might be expressed as 1:1.5, 1.5:1, or simply as 1.5, depending on the meter scale. In practical terms, an SWR of 1.5 is poor because it means that at least 20 percent of the power is being reflected and lost.

SWR can be converted to reflection coefficient by dividing (SWR − 1) by (SWR + 1). For example, using 1.5 SWR, the reflection coefficient is: (1.5 − 1)/(1.5 + 1) = 0.5/2.5 = 0.2 reflection coefficient. In commercial

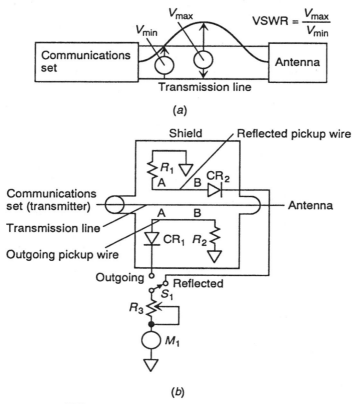

Figure 2.19 SWR measurements.

SWR meters used for RF work, it is not necessary to calculate either the reflection coefficient or SWR. This is done automatically by the SWR meter. (The meter is actually reading the reflection coefficient, but the scale indicates SWR. If you have a reflection coefficient of 0.2, the SWR readout is 1.5.)

Figure 2.19B shows the basic SWR meter circuit. There are two pickup wires, both parallel to the center conductor of the transmission line. Any RF voltage on either of the parallel pickups is rectified and applied to the meter through switch S1. Each pickup wire is terminated in the impedance of the transmission line by corresponding resistors R1 and R2 (typically 50 to 52 Ω).

The outgoing signal (transmitter to antenna) is absorbed by R1, so there is no outgoing voltage on the reflected pickup wire beyond point A. However, the outgoing voltage remains on the transmission line at the outgoing pickup wire. This signal is rectified by CR1 and appears as a reading on the meter when S1 is in the outgoing position.

The opposite occurs for the reflected voltage (antenna to transmitter). No reflected voltage is on the outgoing pickup wire beyond point B because the reflected voltage is absorbed by R2. The reflected voltage does appear on the reflected pickup wire beyond this point and is rectified by CR2. The reflected voltage appears on meter M1 when S1 is in the reflected position.

In use, switch S1 is set to read the outgoing voltage, and resistor R3 is adjusted until the meter needle is aligned with the "set" or "calibrate" line (near the right end of the meter scale). Switch S1 is then set to read the reflected voltage, and then the meter needle moves to the SWR position.

SWR meters often do not read beyond 1:3 because a reading above 1:3 indicates a poor match. Make certain that you understand the scale used on the SWR meter. For example, a typical SWR meter is rated at 1:3, meaning that the scale reads from 1:1 (perfect) to 1:3 (poor). However, the scale indications are 1, 1.5, 2, and 3. These scale indications mean 1:1, 1:1.5, 1:2, and 1:3, respectively. The scale indications between 1 and 1.5 are the most useful because a good antenna system (antenna and lead-in) typically shows 1.1 or 1.2. Anything between 1.2 and 1.5 is on the borderline.

2.18.2 Spectrum analysis

Spectrum analysis is most useful in RF work where FM is involved, although spectrum analyzers can be useful in AM and SSB applications. As shown in Fig. 2.20A, a spectrum analyzer is essentially a narrowband receiver, electrically tuned over a given frequency range, combined with a scope or display tube. The oscillator is switched over a given range of frequencies by a sweep-generator circuit. Because the

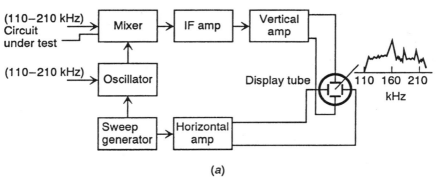

(a)

Figure 2.20 Spectrum analyzer circuits and displays.

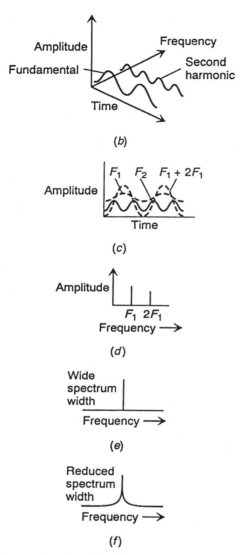

Figure 2.20 *(Continued)*

IF-amplifier passband remains fixed, the input circuits and mixer are swept over a corresponding range of frequencies.

As an example, if the intermediate frequency is 10 kHz and the oscillator sweeps from 100 to 200 kHz, the input is capable of receiving signals in the range of 110 to 210 kHz. The output of the IF amplifier is further amplified and supplied to the vertical deflection plates of the display tube. The horizontal plates receive a signal from the sweep-generator circuit that is used to control the oscillator frequency. As a result, the length of the horizontal sweep represents the total sweep-frequency spectrum (if the sweep is from 110 to 210 kHz, the left end of the display-tube horizontal trace represents 110 kHz and the right-hand end represents 210 kHz). Any point along the horizontal trace represents a corresponding frequency (with the midpoint representing 160 kHz, in this example).

2.18.3 Spectrum Display

Figure 2.20B, 2.20C, and 2.20D shows the relationship of time-amplitude and frequency amplitude displays. A conventional scope produces at time-amplitude display. For example, pulse rise time and width are read directly on the horizontal axis of a scope tube. A spectrum analyzer produces a frequency-amplitude display where signals (unmodulated, AM, FM, pulse, or digital) are broken down into individual components and displayed on a horizontal axis.

In Fig. 2.20B, both the time-amplitude and frequency-amplitude coordinates are shown together. The example given is that showing the addition of a fundamental frequency and second harmonic. In Fig. 2.20C, only the time-amplitude coordinates are shown. The solid line (which is the composite of fundamental F_1 and $2F_1$) is the only display that appears on a conventional scope. In Fig. 2.20D, only the frequency-amplitude coordinates are shown.

2.18.4 Fourier and transform analysis

Spectrum analyzers are often used in conjunction with Fourier and transform analysis. Both of these techniques are quite complex and beyond the scope of this book (and the author). Instead, the practical aspects of spectrum analysis during RF tests are concentrated upon. That is, the book covers what display results from a given input signal and how the display can be interpreted.

2.18.5 Unmodulated spectrum displays

If the spectrum-analyzer oscillator sweeps through an unmodulated or CW signal slowly, the resulting response on the analyzer screen is simply a plot of the analyzer IF-amplifier passband. A pure CW signal has, by definition, energy at only one frequency and should, therefore, appear as a single spike on the analyzer screen (Fig. 2.20E). This occurs provided that the total sweep width (the so-called *spectrum width*) is wide enough, compared to the IF passband of the analyzer. As spectrum width is reduced, the spike response begins to spread out until the IF-passband characteristics begin to appear, as shown in Fig. 2.20F.

2.18.6 AM spectrum displays

A pure sine wave represents a single frequency. The spectrum of a pure sine wave is shown in Fig. 2.21A, and is the same as the unmodulated signal display in Fig. 2.20E (a single vertical line). The height of the line F_0 represents the power contained in the single frequency.

Figure 2.21B shows the spectrum for a single sine-wave frequency (F_0), amplitude modulated by a second sine wave (F_1). In this case, two *sidebands* are formed, one higher and one lower than the frequency (F_0). These sidebands correspond to the sum and difference frequencies, as shown. If more than one modulation frequency is used (as is the case with most practical AM signals), two sidebands are added for each frequency.

When the frequency, spectrum width, and vertical response of the analyzer are properly calibrated, it is possible to find:

1. The carrier frequency.
2. The modulation frequency.
3. The modulation percentage.
4. The nonlinear modulation (if any).
5. Incidental FM (if any).

The rules for interpreting AM spectrum displays are summarized in Fig. 2.21B.

The carrier frequency is determined by the position of the center vertical line (F_0) on the horizontal axis. For example, if the total spectrum is from 100 to 200 kHz, and F_0 is in the center, as shown in Fig. 2.21B, the carrier frequency is 150 kHz.

The modulation frequency is determined by the position of the sideband line ($F_0 - F_1$ or $F_0 + F_1$), on the horizontal axis. For example, if sideband $F_0 - F_1$ is at 140 kHz and F_0 is at 150 kHz (as shown), the modulating frequency is 10 kHz. Under these conditions, the upper sideband ($F_0 + F_1$) should be 160 kHz. The distance between the carrier line (F_1) and either sideband is sometimes known as the *frequency dispersion* and is equal to the modulation frequency.

The modulation percentage is determined by the ratio of the sideband amplitude to the carrier amplitude. The amplitude of either sideband with respect to the carrier amplitude is one half of the percentage of modulation. For example, if the carrier amplitude is 100 mV and either sideband is 50 mV, this indicates 100-percent modulation.

Practical Considerations for RF   75

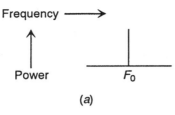

(a)

A = carrier (150 kHz)
Distance between A and B, or A and C = modulation frequency (10 kHz)
Ratio of D to E, or F to E = one-half percent of modulation

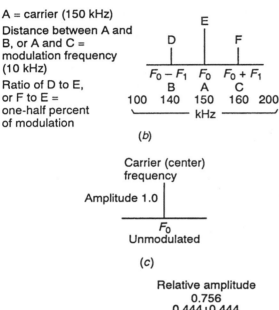

(b)

(c)

Modulation index = 1.0
(deviation of 1 kHz)

(d)

Modulation index = 5.0
(deviation of 5 kHz)

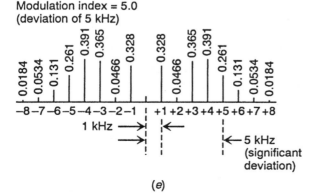

(e)

Figure 2.21 Spectrum analyzer AM/FM displays.

If the carrier amplitude is 100 mV and either sideband is 33 mV, this indicates 66-percent modulation.

Nonlinear modulation is indicated when the sidebands are of unequal amplitude or are not equally spaced on both sides of the carrier frequency. Unequal amplitude indicates nonlinear modulation, which results from a form of undesired frequency modulation combined with amplitude modulation.

Incidental FM is indicated by a shift in the vertical signals along the horizontal axis. For example, any horizontal "jitter" of the signals indicates rapid frequency modulation of the carrier.

In practical tests, carrier signals are often amplitude-modulated at many frequencies simultaneously. This results in many sidebands (two for each modulating frequency) on the display. To resolve this complex spectrum, you must be sure that the analyzer bandwith is less than the lowest modulating frequency or less than the difference between any two modulating frequencies, whichever is smaller.

Overmodulation also produces extra sideband frequencies. The spectrum for overmodulation is very similar to multifrequency modulation. However, overmodulation is usually distinguished from multifrequency modulation by the following:

1. The spacing between overmodulated sidebands is equal, but multifrequency sidebands might be arbitrarily spaced (unless the modulating frequencies are harmonically related).
2. The amplitude of the overmodulated sidebands decreases progressively out from the carrier, but the amplitude of the multifrequency signals is determined by the modulation percentage of each frequency and can be arbitrary.

2.18.7 FM spectrum displays

The mathematical expression for an FM waveform is long and complex, involving a special mathematical operator known as the *Bessel function*. However, the spectrum representation of the FM waveform is relatively simple.

Figure 2.21C shows an unmodulated-carrier spectrum waveform. Figure 2.21D shows the relative amplitudes of the same waveform when the carrier is frequency modulated with a deviation of 1 kHz (modulation index of 1.0). Figure 2.21E shows the relative amplitudes of the waveform when the carrier is frequency modulated with a deviation of 5 kHz (modulation index of 5.0). The modulation index can be found by: *modulation index = maximum frequency deviation ÷ modulation frequency*.

The term *maximum frequency deviation* is theoretical. If a CW signal (F_c) is frequency modulated at a rate F_r, an infinite number of sidebands results. These sidebands are located at intervals of $F_c N$, where

N = 1, 2, 3, etc. However, as a practical matter, only the sidebands containing significant power are usually considered. For a quick approximation of the bandwidth occupied by the significant sidebands, multiply the sum of the carrier deviation and the modulating frequency by 2, or: *bandwidth* = 2 (*carrier deviation* + *modulating frequency*).

As a guideline, when using a spectrum analyzer to find the maximum deviation of an FM signal, locate the sideband where the amplitude begins to drop and continues to drop when the frequency moves from the center. For example, in Fig. 2.21E, sidebands 1, 2, 3, and 4 rise and fall, but sideband 5 falls, and all sidebands after 5 continue to fall. Because each sideband is 1 kHz from the center, this indicates a practical or significant deviation of 5 kHz. (It also indicates a modulation index of 5.0, in this case.)

As in the case of AM, the center and modulation frequencies for FM can be determined with the spectrum analyzer.

The FM carrier frequency is determined by the position of the center vertical line on the horizontal axis. (The centerline is now always the highest amplitude, as shown in Fig. 2.21E.)

The FM modulating frequency is determined by the position of the sidebands in relation to the centerline or by the distance between sidebands (frequency dispersion).

2.19 Basic RF Measurements

This section covers a few basic test and measurement procedures for components associated with wireless/RF ICs. The procedures are particularly useful during design or experimentation. Typically, the tests should be made when the circuit is first completed in experimental form. If the test results are not as desired, the component values should be changed (as necessary) to get the desired results. Also, RF circuits should always be retested in final form (with all components soldered in place). This shows if the circuit characteristics change because of the physical relocation of components.

Although this procedure might seem unnecessary, it is especially important at higher radio frequencies. Often, capacitance or inductance exists between components, from components to wiring or board traces, and between traces. These stray "components" can add to the reactance and impedance of circuit components. When the physical location of parts is changed, the stray reactances change and alter circuit performance.

2.19.1 RF probes

An RF probe is required when the voltages or signals to be measured are beyond the frequency capability of the meter or scope. Such probes rectify the RF signals into a dc output, which is almost equal to the peak

RF voltage. The dc output of the probe is then applied to the meter or scope input and is displayed as a voltage readout in the normal manner. If a probe is available as an accessory for a particular meter or scope, that probe should be used in favor of any experimental or homemade probe. The manufacturer's probe is matched to the meter/scope in voltage calibration, frequency compensation, etc.

2.19.2 Resonant-frequency measurements

As shown in Fig. 2.22, a meter can be used in conjunction with a signal generator to find the resonant frequency of either series or parallel LC circuits. As covered, the inductive (L) and capacitive (C) reactances are equal at the resonant frequency. A parallel LC circuit acts as a high impedance at the resonant frequency; a series LC circuit has low impedance at resonance. The following procedure can be applied to any type of LC circuit.

1. Connect the equipment, as shown in Fig. 2.22. Use the connections in Fig. 2.22A for parallel-resonant LC circuits, or the connections in Fig. 2.22B for series-resonant circuits.

2. Adjust the generator output until a convenient midscale indication is obtained on the meter. Use an unmodulated signal output from the generator.

3. Starting at a frequency well below the lowest possible frequency of the circuit under test, slowly increase the generator output frequency. If you can not judge the approximate resonant frequency, use the lowest generator frequency.

4. If the circuit being tested is parallel resonant, watch the meter for a maximum (peak) indication.

5. If the circuit being tested is series resonant, watch the meter for a minimum (dip) indication.

6. The resonant frequency of the circuit under test is the one at which a maximum (for parallel) or minimum (for series) indication is on the meter.

7. Peak or dip indications might be at harmonics of the resonant frequency. The test is most efficient when the approximate resonant frequency is known.

8. The value of load resistance R_L is not crucial. The load is shunted across the LC circuit to flatten, or broaden, the resonant response (to lower the circuit Q), causing the voltage maximum or minimum to be approached more slowly. A suitable trial value for R_L is 100 kΩ. A lower R_L sharpens the resonant response, but a higher value flattens the curve.

2.19.3 Coil inductance measurements

As shown in Fig. 2.23, a meter can be used in conjunction with a signal generator and fixed capacitor (of known value and accuracy) to find the inductance of a coil.

1. Connect the equipment as shown in Fig. 2.23. Use a capacitance value such as 10 μF, 100 pF, or some other even number to simplify the calculation.

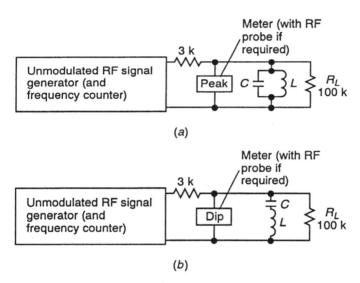

Figure 2.22 Basic RF voltage measurement.

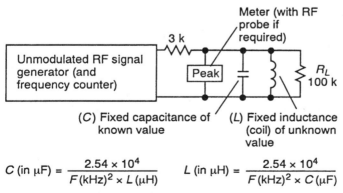

$$C \text{ (in } \mu F) = \frac{2.54 \times 10^4}{F(\text{kHz})^2 \times L(\mu H)} \qquad L \text{ (in } \mu H) = \frac{2.54 \times 10^4}{F(\text{kHz})^2 \times C(\mu F)}$$

Figure 2.23 Basic coil inductance measurements.

2. Adjust the generator output until a convenient midscale indication is obtained on the meter. Use an unmodulated signal output from the generator.

3. Starting at a frequency well below the lowest possible resonant frequency of the inductance/capacitance combination under test, slowly increase the generator frequency. If you can not judge the approximate resonant frequency, use the lowest generator frequency.

4. Watch the meter for a maximum (peak) indication. Note the frequency at which the peak indication occurs. This is the resonant frequency of the circuit.

5. Using the resonant frequency and the known capacitance value, calculate the unknown inductance using the equation in Fig. 4.23.

6. The procedure can be reversed to find an unknown capacitance value, when a known inductance value is available.

2.19.4 Self-resonance and distributed-capacitance measurements

As shown in Fig. 2.24, a meter can be used in conjunction with a signal generator to find both the self-resonant frequency and the distributed capacitance of a coil. No matter what design or winding method is used, some distributed capacitance is in any coil. When this capacitance combines with the coil inductance, a resonant circuit is formed. The resonant frequency is usually quite high in relation to the frequency at which the coil is used. However, because self-resonance might be at or near a harmonic of the frequency to be used, the self-resonant effect can limit the usefulness of the coil in LC circuits. Some coils, particularly RF chokes, might have more than one self-resonant frequency.

1. Connect the equipment, as shown in Fig. 2.24.

2. Adjust the generator output amplitude until a convenient midscale indication is obtained on the meter. Use an unmodulation signal output from the generator.

3. Tune the generator over the entire frequency range, starting at the lowest frequency. Watch the meter for either peak or dip indications. A peak or dip indicates that the coil is at a self-resonant point. The generator output frequency at that point is the self-resonant frequency. Be certain that peak or dip indications are not the result of changes in generator output level. Even the best laboratory generators might not produce a flat (constant level) output over the entire frequency range.

4. Because there might be more than one self-resonant point, tune through the entire signal-generator range. Try to cover a frequency

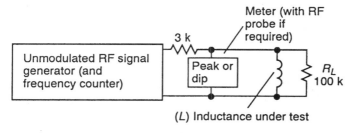

$$C \text{ (distributed capacitance in pF)} = \frac{2.54 \times 10^4}{F(\text{MHz})^2 \times L(\mu H)}$$

Figure 2.24 Basic coil self-resonance and distributed-capacitance measurements.

range up to at least the third harmonic of the highest frequency involved in a resonant-circuit design.

5. When the resonant frequency is found, calculate the distributed capacitance using the equation in Fig. 2.24. For example, assume that a coil with an inductance of 7 µH is found to be self-resonant at 50 MHz; C (distributed capacitance) = $(2.54 \times 10^4)/(50^2 \times 7)$ = 1.43 pF.

2.19.5 Resonant-circuit Q measurements

Figure 2.25 shows how the Q of a tuned RF circuit can be measured. As covered in Section 2.2, a resonant circuit with a high Q produces a sharp resonance curve (narrow bandwidth); a low Q produces a broad resonance curve or wide bandwidth.

Figure 2.25A shows the test circuit in which the signal generator is connected directly to the input of a complete stage. Figure 2-25B shows the indirect method of connecting the signal generator to the input. When the stage or circuit has sufficient gain to provide a good reading on the meter with a nominal output from the generator, the indirect method (with isolating resistor) is preferred. Any generator has some output impedance (typically 50 Ω). When this resistance is connected directly to the tuned circuit, the Q is lowered, and the response becomes broader. (In some cases, the generator output seriously detunes the circuit.)

Figure 2.25C shows the test circuit for a single component (such as an IF transformer). The value of the isolating resistance is not crucial and is typically in the range of 100 kΩ. The procedure for determining Q using any of the circuits in Fig. 2.25 is:

1. Connect the equipment, as shown in Fig. 2.25. Notice that a load is shown in Fig. 2.25C. When a circuit is normally used with a load, the most realistic Q measurement is made with the circuit terminated

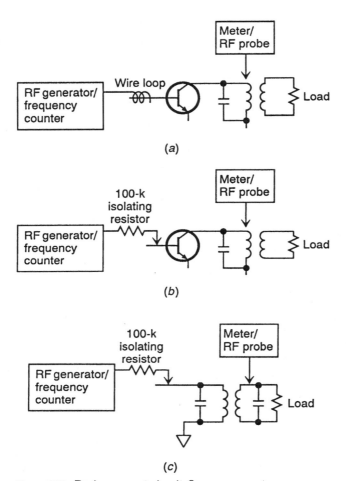

Figure 2.25 Basic resonant-circuit Q measurements.

in that load value. A fixed resistance can be used to simulate the load. (The Q of a resonant circuit often depends on the load value.)

2. Tune the generator to the circuit resonant frequency. Use an unmodulated output from the generator.
3. Tune the generator frequency for maximum reading on the meter. Notice the generator frequency.
4. Tune the generator below resonance until the meter reading is 0.707 times the maximum reading. Notice the generator frequency. To make the calculation more convenient, adjust the generator output level so that the meter reading is some even value, such as 1 V

or 10 V, after the generator is tuned for maximum. This makes it easy to find the 0.707 mark.

5. Tune the generator above resonance until the meter reading is 0.707 times the maximum reading. Notice the generator frequency.
6. Calculate the resonant-circuit Q using the equations in Fig. 2.2. For example, assume that the maximum meter indication occurs at 455 kHz (F_R), the below-resonance indication is 453 kHz (F_2), and the above-resonance indication is 458 kHz (F_1). Then $Q = 455/(458 - 453) = 91$.

2.19.6 Resonant-circuit impedance measurements

Figure 2.26 shows how the impedance of a resonant circuit can be measured. Any resonant circuit has some impedance at the resonant frequency. The impedance changes as frequency changes. This includes transformers (tuned and untuned), RF tank circuits, etc. In theory, a series-resonant circuit has zero impedance and a parallel-resonant circuit has infinite impedance at the resonant frequency. In practical RF circuits, this is impossible because some resistance is always in the circuit.

It is often convenient to find the impedance of an experimental resonant circuit at a given frequency. Also, it might be necessary to find the impedance of a component in an experimental circuit so that other circuit values can be designed around the impedance. For example, an IF transformer presents an impedance at both the primary and secondary windings. These values might not be specified on the transformer datasheet.

If the circuit or component under measurement has both an input and output (such as a transformer), the opposite side or winding must be terminated in the normal load, as shown. If the impedance of a tuned circuit is to be measured, tune the circuit to peak or dip, and then measure the impedance at resonance. When the resonant impedance is found,

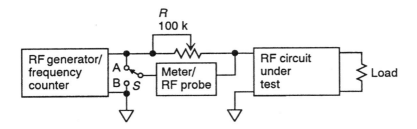

Figure 2.26 Basic resonant-circuit impedance measurements.

the generator can be tuned to other frequencies to find the corresponding impedance (if required). Notice that a high-impedance digital meter provides the least loading effect on the circuit and thus produces the most accurate indication.

Adjust the generator to the frequency (or frequencies) at which impedance is to be measured. Move switch S back and forth between positions A and B while adjusting resistance R until the voltage reading is the same in both positions of the switch. Disconnect resistor R from the circuit, and measure the dc resistance of R with an ohmmeter. The dc resistance of R is then equal to the impedance at the circuit input.

The accuracy of the impedance measurement depends on the accuracy with which the dc resistance is measured. A noninductive resistance must be used. The impedance found by this method applies only to the frequency used during the test.

Chapter 3

Direct-Conversion Tuners for Digital DBS

This chapter is devoted to optimizing direct-conversion tuner ICs for digital DBS applications. The MAX2102 is selected as an example. Such tuner ICs are used in digital direct-broadcast satellite (DBS) television set-top box units to provide direct conversion of TV signals instead of IF-based conversion.

3.1 Circuit Description for MAX2102

Figures 3.1, 3.2, 3.3, and 3.4 show the functional diagram, typical application circuit, typical external connections, and pin descriptions, respectively. The MAX2102 directly tunes L-band (see Fig. 1.1) signals to the baseband using a broadband I/Q downconverter. The operating frequency range spans from at least 950 MHz to 2150 MHz. As shown in Fig. 3.1, the IC includes LNA (low-noise amplifier) with AGC, two downconverter mixers, an oscillator-buffer with 90° quadrature generation and a prescaler, and baseband amplifiers. For readers not familiar with the terms *quadrature, I, Q,* or *baseband,* read *Lenk's Television Handbook* (McGraw-Hill, 1995).

The front-end AGC dynamic range is greater than 50 dB. Specifically, the AGC control can be adjusted so that a sine wave at RFIN ranging in power from -69 dBm to -19 dBm will produce a sine wave at I_{OUT} and Q_{OUT} at 500-mVp-p levels. The noise figure is lowest when the AGC is at its maximum gain setting. The VSWR (see Section 2.18.1) at RFIN is unaffected by the AGC setting.

The local-oscillator (LO) buffer accepts an external local-oscillator signal at LO, $\overline{LO}$, and internally limits the signals to provide a consistent

86 Chapter Three

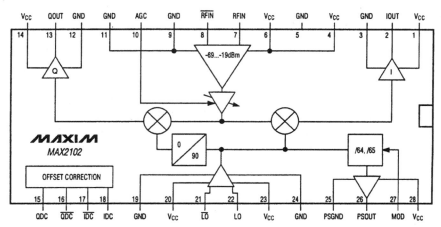

Figure 3.1 MAX2102 functional diagram (*Maxim,* New Releases Data Book, *1998, p. 10–7*).

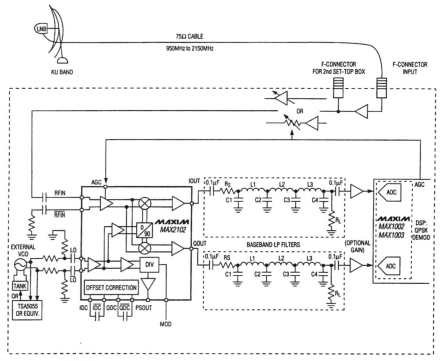

Figure 3.2 MAX2102 typical application (*Maxim,* New Releases Data Book, *1998, p. 10–16*).

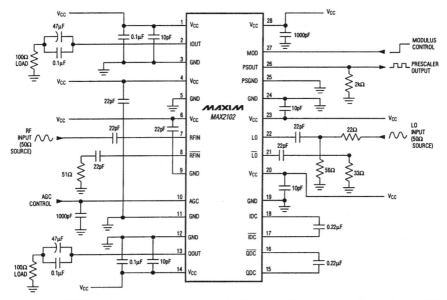

Figure 3.3 MAX2102 typical operating circuit (*Maxim*, New Releases Data Book, *1998*, p. 10–15).

on-chip LO level. The LO input drive level should be maintained with the specified limits (as covered in Section 3.3).

The quadrature downconverter follows the front-end AGC. Two mixers are driven by the previous stage AGC amplifier output. The mixer LO ports are fed with the two LO signals, which are 90° apart in phase. These quadrature LO signals are generated on-chip using the LO signals from the LO buffer.

The resulting I/Q baseband signals are fed through separate I and Q channel baseband amplifiers, and appear at I_{OUT} and Q_{OUT}. The outputs are capable of driving low-pass filters with 100-Ω characteristic impedance (that is, the equivalent of an ac-coupled 100-Ω load). The baseband -3-dB output bandwidth is over 90 MHz.

3.2 Front-End Tuner Circuitry for DBS Tuners

In a typical application, the signal path ahead of the MAX2102 includes a discrete LNA/buffer and a PIN-diode attenuator, as shown in Fig. 3.2. As an alternate, a dual-gate GaAsFET can serve this function. The circuitry is usually required to meet system

PIN	NAME	FUNCTION
1	V_{CC}	Baseband +5V Supply. Bypass with a 10pF capacitor from this pin to pin 3 (GND), as close to the IC as possible. Connect an additional 0.1μF capacitor in parallel with the 10pF capacitor (placement less critical).
2	IOUT	I Channel Baseband Output
3, 12	GND	Baseband Ground
4	V_{CC}	RF +5V Supply. Bypass with a 22pF capacitor from this pin to pin 11 (GND), as close to the IC as possible.
5	GND	Ground (substrate)
6	V_{CC}	RF +5V Supply. Bypass with a 22pF capacitor from this pin to pin 9 (GND), as close to the IC as possible.
7	RFIN	RF Noninverting Input. Couple through a 22pF capacitor directly to a 50Ω signal source.
8	$\overline{RFIN}$	RF Inverting Input. Connect to a 22pF series capacitor and a 51Ω resistor to ground.
9, 11, 19, 24	GND	RF Ground. Connect directly to the ground plane.
10	AGC	Automatic Gain-Control Input. Bypass this pin with a 1000pF capacitor close to the pin, to minimize coupling.
13	QOUT	Q Channel Baseband Output
14	V_{CC}	Baseband +5V Supply. Bypass with a 10pF capacitor from this pin to pin 12 (GND), as close to the IC as possible. Connect an additional 0.1μF capacitor in parallel with the 10pF capacitor (placement less critical).
15	QDC	Q Channel Offset-Correction Noninverting Input. Connect a 0.22μF (typ) capacitor between QDC and $\overline{QDC}$. This capacitor must be placed as close to the IC as possible. See *Layout Considerations* section.
16	$\overline{QDC}$	Q Channel Offset-Correction Inverting Input. Connect a 0.22μF (typ) capacitor between QDC and $\overline{QDC}$. This capacitor must be placed as close to the IC as possible. See *Layout Considerations* section.
17	$\overline{IDC}$	I Channel Offset-Correction Inverting Input. Connect a 0.22μF (typ) capacitor between IDC and $\overline{IDC}$. This capacitor must be placed as close to the IC as possible. See *Layout Considerations* section.
18	IDC	I Channel Offset-Correction Noninverting Input. Connect a 0.22μF (typ) capacitor between IDC and $\overline{IDC}$. This capacitor must be placed as close to the IC as possible. See *Layout Considerations* section.
20	V_{CC}	RF +5V Supply. Bypass with a 10pF capacitor from this pin to pin 19 (GND) as close to the IC as possible.
21	$\overline{LO}$	Local-Oscillator Complementary Input Port (Figure 1)
22	LO	Local-Oscillator Input Port (Figure 1)
23	V_{CC}	RF +5V Supply. Bypass with a 10pF capacitor from this pin to pin 24 (GND), as close to the IC as possible.
25	PSGND	Prescaler Ground. To disable the prescaler, leave this pin open.
26	PSOUT	Prescaler Output. Drives CMOS load. Connect 2kΩ from this pin to GND (if the prescaler is enabled).
27	MOD	Prescaler Modulus Control. Leave open when the prescaler is disabled.
28	V_{CC}	Prescaler +5V Supply. Bypass with a 1000pF capacitor. **Must be connected even if the prescaler is disabled.**

Figure 3.4 MAX2102 pin description (*Maxim*, New Releases Data Book, *1998, p. 10–14*).

noise-figure limitations and might provide a buffered F-connector output. The circuit might also be required to meet some stringent LO-leakage specifications. The PIN attenuator is typically controlled by the same voltage as the MAX2102 AGC control pin so that a single AGC line from the baseband processor can control the entire tuner.

In some applications, a varactor-tuned preselection bandpass filter is added between the discrete LNA/buffer and the MAX2102. This is usually required only for very high-linearity tuners, such as those designed for low-data-rate applications. The filter provides a means of broadly filtering adjacent interference. This improves the intermodulation performance of the tuner. Also, the filter removes the RF interference at twice the LO frequency, which would otherwise add to the co-channel interference (the MAX2102 alone rejects this carrier to approximately 32.3 dBc).

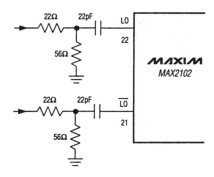

Figure 3.5 Connections for differential LO drive (*Maxim*, New Releases Data Book, *1998, p. 10–17*).

LO POWER:
-15dBm to -5dBm,
POWER INTO LO, $\overline{LO}$.
LO, $\overline{LO}$ DRIVEN DIFFERENTIALLY

3.3 External Oscillator Requirements

Because the MAX2102 is a direct-conversion receiver, the external LO must tune to the same frequencies as the desired RF input signals. The input ports (LO, $\overline{LO}$) accept either a single-ended or differential (balanced) LO signal. A differential LO offers reduced LO leakage to the RFIN port, as well as lower spurious downconversion levels of RF signals (which are at twice the LO frequency). For optimum performance, be certain that the LO and $\overline{LO}$ traces are symmetrical. Figure 3.5 shows the connections for a differential LO drive.

The LO drive signals must be maintained to within the specified limits (−15 dBm to −5 dBm). If the LO drive falls below this range, quadrature performance might be affected. Driving LO above the specified limits will cause LO leakage level at RFIN (which might be acceptable in some applications). The MAX2102 offset-correction loop can withstand LO leakage levels corresponding to at least 0 dBm of LO input power drive.

3.4 Prescaler Requirements

Typical stand-alone tuner applications will not use the MAX2102 prescaler functions. Instead, a commercial synthesizer IC, such as the Philips TSA5055, will be used. (The Philips TSA5055 has an internal prescaler.)

To disable the prescaler, disconnect the PSGND pin (leave the pin open). The prescaler will cause an output spur in the baseband spectrum, to a level of about −20 dBc (referred to 500-mVp-p baseband output level). This prescaler output might land within the desired signal bandwidth in some applications, so it must be disabled when not required.

To use the MAX2102 prescaler, connect the PSGND pin to ground. In some applications, the prescaler can be toggled on and off using a MOSFET to switch PSGND to ground. PSGND should be forced to within 100 mV of ground, and the MOSFET must be capable of sinking 15 mA.

PSOUT is capable of driving a typical CMOS load of 10 kΩ in parallel with 5 pF. A 2-kΩ pull-down resistor must be connected from PSOUT to GND.

The prescaler requires a stable level at the MOD pin 12 ns before the falling edge of PREOUT to assert the desired modulus. The level at MOD must remain static until 3 ns after the falling edge. See the notes on Fig. 3.4, pins 25 through 28, for optimizing prescaler functions.

3.5 Baseband Amplifiers

The MAX2102 baseband amplifiers provide more than 2 Vp-p of swing at I_{OUT} and Q_{OUT}, and are capable of driving 100 Ω. I_{OUT} and Q_{OUT} must be ac coupled to any low-pass filters (such as those shown in Fig. 3.2).

In a typical application, I_{OUT} and Q_{OUT} drive a 5th or 7th-order low-pass filter for ADC antialiasing purposes. This is covered further in Section 3.9. In some cases, additional gain might be required after the filters. The gain can be provided by a pair of video-speed op amps, such as the MAX4216 dual video op amp. As an alternate, the MAX1002/MAX1003 dual ADC has built-in gain ahead of the ADCs; it is capable of digitizing levels as low as 125 mVp-p.

3.6 Offset Correction

The internal offset-correction amplifiers remove the dc offsets present in the baseband amplifiers. The offset-correction loop effectively ac couples the baseband signal path, yielding a -3-dB high-pass corner frequency as follows:

$$f(-3 \text{ dB}) = 100/C_{DC} (\mu F)$$

where C_{DC} is the value of the capacitors (in μF), in QDC, $\overline{QDC}$ and IDC, $\overline{IDC}$.

For applications where the dc information must be maintained through the signal path, the offset correction can be disabled by connecting QDC, $\overline{QDC}$ and IDC, $\overline{IDC}$ directly to ground. Disabling the offset correction will effectively limit the input dynamic range of the MAX2102. Typical input dynamic range will be about -35 dBm to -19 dBm for a single-ended LO drive, and -55 dBm to -19 dBm for a differential LO drive (Fig. 3.5).

3.7 Optimizing MAX2102 Layout

Although a ground plane is essential for any configuration, remove the ground plane under the external VCO area to reduce parasitic capacitance. If a ground plane is used under the low-pass filters, the filter shape might be slightly offset because of parasitic capacitance.

In a direct-conversion receiver, LO leakage to the RF input connector is a major concern because filtering of the LO is impossible (the LO operates at the same frequency as the RF input). The external VCO section should be housed in a separate shielded compartment (if practical).

A differential or balanced LO (such as shown in Fig. 3.5) will dramatically reduce LO leakage. Also, use of a coplanar waveguide transmission line will reduce LO leakage. The evaluation kit for the MAX2102 uses the coplanar waveguide.

Observe the power-supply bypass-capacitor connection notes in Fig. 3.4, particularly for pins 1, 3, 4, 6, 9, 11, 12, 14, 19, 20, 23, and 34. Traces from these IC pins to the bypass capacitors must be kept to an absolute minimum. Where possible, make these connections on the top side of the board.

Minimize parasitic capacitance to ground around the offset-correction circuit (pins 15 to 18) by removing the ground plane beneath these pins and placing the offset-correction capacitors as close to the IC as possible.

The MAX2102 evaluation kit includes ferrite beads in series with power-supply leads. The beads might not be required for all applications.

3.8 Power-Supply Sequencing

The MAX2102 has several +5-V supply pins. The supply layout should be in the star format (where all pins tie to the same central point), with a bypass capacitor that dominates the rise time of the supply at the center of the star. This will ensure that all pins "see" approximately the same voltage during power up.

The prescaler V_{CC} (pin 28) must be connected to the same V_{CC} as the other V_{CC} pins—even if the prescaler is not used. Leaving PSGND open will still disable the prescaler function, and the prescaler will not dissipate any power.

3.9 Low-Pass Filters in Direct-Conversion Tuners

Figures 3.6 and 3.7 show the basic filter requirements. Figure 3.8 shows the suggested component values for the antialiasing low-pass filters shown in Fig. 3.2. The values given are for a Chebyshev filter with 0.1-dB ripple. For readers not familiar with Chebyshev or other filters, read *Simplified Design of Filter Circuits* (Butterworth-Heinemann, Newnes, 1999).

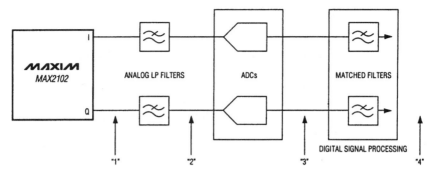

Figure 3.6 I and Q signal paths (*Maxim,* New Releases Data Book, *1998, p. 10–19*).

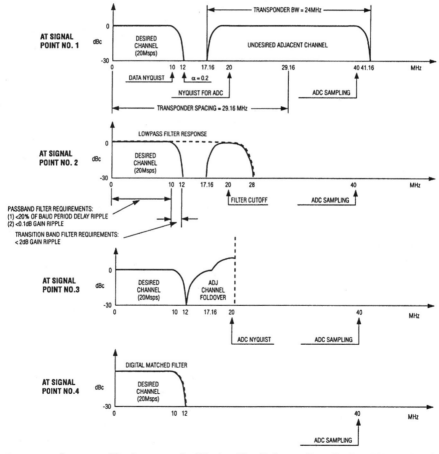

Figure 3.7 Low-pass filtering example (*Maxim,* New Releases Data Book, *1998, p. 10–20*).

ADC SAMPLING RATE (Msps)	FILTER TYPE	Rs (Ω)	C1 (pF)	L1 (nH)	C2 (pF)	L2 (nH)	C3 (pF)	L3 (nH)	C4 (pF)	R_L (kΩ)
40	0.1dB Chebyshev, f_C = 20MHz	100	39	910	120	1500	150	1500	120	10
60	0.1dB Chebyshev, f_C = 30MHz	100	22	620	82	910	100	1000	820	10
90	0.1dB Chebyshev, f_C = 45MHz	100	18	390	56	620	68	680	56	10

Note: Suggested types: inductors: Coilcraft 1008CS, tolerance = ±5%; capacitors: use tolerance = ±2%. Refer to Figure 2 for circuit diagram.

Figure 3.8 Suggested component values for antialiasing low-pass filters (*Maxim*, New Releases Data Book, *1998, p. 10–18*).

Chapter

4

IF Transceiver with Limiter and RSSI

This chapter is devoted to optimizing IF transceiver ICs. The MAX2511 is selected as an example. Such ICs are complete, highly integrated IF transceivers (combined transmitter and receiver) used in applications that require dual-conversion (such as wireless handsets and base stations, or wireless data links). The MAX2511 can also be used as a single-conversion transceiver if the RF operating frequency ranges from 200 MHz to 440 MHz.

4.1 Circuit Description for MAX2511

Figures 4.1, 4.2, and 4.3 show the functional diagram, pin descriptions, and typical operating circuit, respectively. In a typical application, the receiver downconverts a high IF/RF (200 MHz to 440 MHz) to a 10.7-MHz low IF using an image-rejection mixer. Functions include an image-rejection downconverter with 34 dB of image suppression, followed by an IF buffer that can drive an off-chip IF filter; an on-chip limiting amplifier that offers 90 dB of monotonic RSSI (received-signal-strength indicator); and a limiter output driver.

The transmit image-rejection mixer generates a clean output spectrum to minimize filter requirements. It is followed by a 40-dB VGA (variable-gain amplifier) that maintains intermodulation levels (IM3) below −35 dBc. The maximum output power is 2 dBm. A VCO and oscillator-buffer for driving an external prescaler are also included. The MAX2511 operates from a 2.7-V to 5.5-V supply and includes flexible power-management control. Supply current is reduced to 0.1 μA in the shutdown mode.

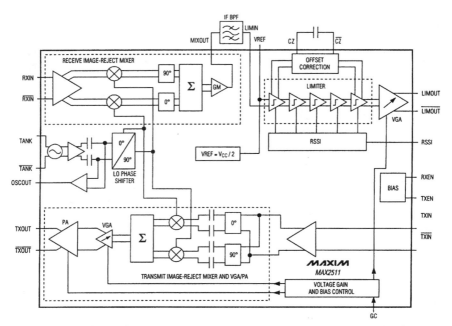

Figure 4.1 MAX2511 functional diagram (*Maxim*, New Releases Data Book, *1998*, p. 10–42).

For applications that require in-phase (I) and quadrature (Q) baseband architecture for the transmitter, Maxim offers the corresponding MAX2510 transceiver. As shown in Fig. 4.4, the MAX2510 has features similar to those of the MAX2511, but it upconverts I/Q baseband signals using a quadrature upconverter.

4.2 Receiver Circuit

As shown in Fig. 4.1, the receiver consists of two basic blocks: the image-rejection downconverter-mixer and the limiter/RSSI section. The receiver inputs are the RXIN and $\overline{\text{RXIN}}$ pins, which should be ac coupled and might required a matching network, as shown in Fig. 4.3. (Calculations for optimizing the matching network are covered in Section 4.8.)

4.2.1 Image-rejection mixer

The downconverter uses an image-rejection mixer, consisting of an input buffer with dual outputs—each of which is fed to a double-balanced mixer. The LO signal is generated by an on-chip oscillator

IF Transceiver with Limiter and RSSI

PIN	NAME	FUNCTION
1	LIMIN	Limiter Input. Connect a 330Ω (typ) resistor to VREF for DC bias, as shown in the *Typical Operating Circuit*.
2, 3	CZ, $\overline{CZ}$	Offset-Correction Capacitor pins. Connect a 0.01µF capacitor between CZ and $\overline{CZ}$.
4	RSSI	Receive-Signal-Strength-Indicator Output. The voltage on RSSI is proportional to the signal power at LIMIN. The RSSI output sources current pulses into an external capacitor (100pF typ). The output is internally terminated with 6kΩ, and this RC time constant sets the decay time.
5	GC	Gain-Control pin in transmit mode. Applying a DC voltage to GC between 0V and 2.0V adjusts the transmitter gain by 40dB. In receive mode, GC adjusts the limiter output level from 0Vp-p to about 1Vp-p. This pin's input impedance is typically 80kΩ terminated to 1.35V.
6, 9	TANK, $\overline{TANK}$	Tank pins. Connect the resonant tank across these pins, as shown in the *Typical Operating Circuit*.
7, 10	GND	Ground. Connect GND to the PC board ground plane with minimal inductance.
8, 11	V_{CC}	Supply Voltage. Bypass V_{CC} directly to GND. See the *Layout Issues* section.
12	OSCOUT	Oscillator-Buffer Output. OSCOUT provides a buffered oscillator signal (at the oscillator frequency) for driving an external prescaler. This pin is a current output and must be AC-coupled to a resistive load. The output power is typically -9dBm into a 50Ω load. If a larger output swing is required, a larger load resistance (up to 100Ω) can be used.
13, 14	LIMOUT, $\overline{LIMOUT}$	Differential Outputs of the Limiting Amplifier. LIMOUT and $\overline{LIMOUT}$ are open-collector outputs that are internally pulled up to V_{CC} through 1kΩ resistors.
15, 16	$\overline{TXIN}$, TXIN	Differential Inputs of the Image-Reject Upconverter Mixer. $\overline{TXIN}$ and TXIN are high impedance and must be pulled up to V_{CC} through two external resistors whose value is equal to the desired terminating impedance (50Ω to 50kΩ).
17	RXEN	Receiver-Enable pin. When high, RXEN enables the receiver if TXEN is low. If both RXEN and TXEN are high, the part is in standby mode; if both are low, the part is in shutdown. See the *Power Management* section for more details.
18	TXEN	Transmitter-Enable pin. When high, TXEN enables the transmitter, if RXEN is low. If both TXEN and RXEN are high, the part is in standby mode; if both are low, the part is in shutdown. See the *Power Management* section for more details.
19, 21	V_{CC}	Bias V_{CC} Supply pins. Decouple these pins to GND. See the *Layout Issues* section.
20	GND	Receiver/Transmitter Ground pin. Connect to the PC board ground plane with minimal inductance.
22, 25	$\overline{RXIN}$, RXIN	Differential Inputs of the Image-Reject Downconverter Mixer. In most applications, an impedance matching network is required. See the *Applications Information* section for more details.
23, 24	$\overline{TXOUT}$, TXOUT	Differential Outputs of the Image-Reject Upconverter. $\overline{TXOUT}$ and TXOUT must be pulled up to V_{CC} with two external inductors and AC coupled to the load.
26	GND	Receiver Front-End Ground. Connect GND to the PC board ground plane with minimal inductance.
27	MIXOUT	Single-Ended Output of the Image-Reject Downconverter. MIXOUT is high impedance and must be biased to VREF through an external terminating resistor whose value depends on the interstage filter characteristics. See the *Applications Information* section for more details.
28	VREF	Reference Voltage pin. VREF is used to provide an external bias voltage for the MIXOUT and LIMIN pins. Bypass this pin with a 0.1µF capacitor to ground. VREF voltage is equal to V_{CC} / 2. See the *Typical Operating Circuit* for more information.

Figure 4.2 MAX2511 pin descriptions (*Maxim*, New Releases Data Book, *1998, p. 10–41*).

and an external tank circuit. The buffered oscillator signal drives a quadrature phase generator that provides two outputs with 90° of phase shift between them. This pair of LO signals is fed to the two receiver mixers. The mixer outputs are then passed through a pair of phase shifters, which provide 90° of phase shift across their outputs.

The two signals (with phase shift) are summed together, with the final phase relationship such that the desired signal is reinforced and the image signal is largely canceled. The downconverter-mixer output is buffered and converted to a single-ended current output at the MIXOUT pin. MIXOUT can drive a shunt-terminated (330-Ω, 165-Ω load) bandpass filter over a large dynamic range at more than 2 V over the entire supply range.

98   Chapter Four

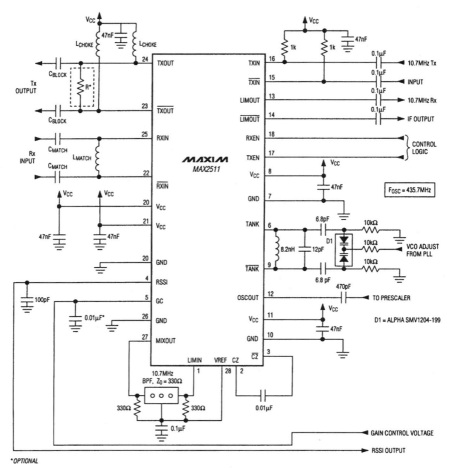

Figure 4.3 MAX2511 typical operating circuit (*Maxim,* New Releases Data Book, *1998,* p. 10–48).

4.2.2 Limiter

The signal from the external bandpass filter is applied to the limiter through LIMIN. This input is centered around the V_{REF} pin voltage. The open-circuit input impedance is typically greater than 10 kΩ, terminated to V_{REF}. For optimum performance, LIMIN must be tied to V_{REF} through the filter terminating impedance (not more than 1 kΩ). The limiter provides a constant output level, which is largely independent of the limiter input-signal level over an 80-dB input range.

The VGA following the limiter provides for easy interfacing of the limiter output to downstream circuitry. The VGA adjusts the limiter-

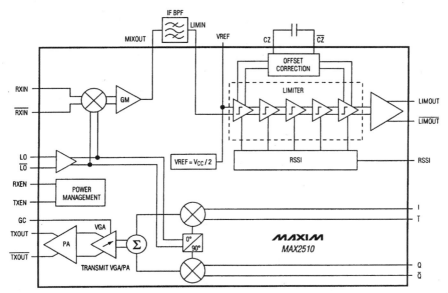

Figure 4.4 MAX2510 typical application circuit (*Maxim,* New Releases Data Book, *1998,* p. 10–32).

output level from 0 V p-p to 1 V p-p when the GC pin is adjusted from 0.5 V to 2.0 V. Using this feature allows the downstream circuits, such as an analog-to-digital converter (ADC), to run at optimum performance by steering the limiter output level to match the desired ADC input level. GC is also used for the transmit (Tx) gain adjustment in the Tx mode (discussed in Section 4.5). It is essential that the GC voltage be kept within an appropriate value for both Tx and Rx (receive) modes.

4.2.3 RSSI

The RSSI output provides a linear indication of the received power level on the LIMIN input. The RSSI monotonic dynamic range exceeds 90 dB and provides better than 80 dB of linear range. The RSSI output is internally terminated with 6 kΩ to GND. This RC time constant sets the decay time for the RSSI indication.

4.3 Transmitter Circuit

As shown in Fig. 4.1, the transmitter image-rejection upconverter-mixer operates in a fashion similar to that of the downconverter-mixer. The transmitter mixer consists of an input buffer-amplifier that drives on-chip IF phase shifters. The shifted signals are then input to a pair of double-balanced mixers, which are driven with the same quadrature (Q) LO source used by the receiver. The mixer outputs are summed together, largely canceling the image signal. The image-canceled signal from the mixer outputs is fed through a VGA with 40 dB of gain-adjust range.

The VGA output is connected to a driver amplifier with an output 1-dB compression point of 2 dBm. The output power can be adjusted from about 2 dBm to less than −40 dBm by controlling the GC pin. The resulting signal appears as a differential output on TXOUT and $\overline{\text{TXOUT}}$ (normally terminated in a 100-Ω differential load).

TXOUT and $\overline{\text{TXOUT}}$ are open-collector outputs and require external pull-up inductors to V_{CC} for proper operation. The outputs also need a dc block so that the load does not affect dc biasing. A shunt resistor across TXOUT and $\overline{\text{TXOUT}}$ can be used to back-terminate an external filter, as shown in Fig. 4.3. It is possible to use the receiver inputs RXIN and $\overline{\text{RXIN}}$ to provide this termination, as described in Section 4.9. For single-ended operation, tie the unused input to V_{CC}.

4.4 Local Oscillator and Oscillator Buffer

As shown in Fig. 4.3, the on-chip LO requires an external LC tank circuit connected across TANK and $\overline{\text{TANK}}$ (calculations for optimizing the tank circuit are covered in Section 4.7). Typically, a dual varactor diode is used to adjust the LO frequency in response to signals from a PLL (phase-locked loop). For optimum performance (lowest phase noise), keep the tank-circuit Q as high as possible. The tank PC board layout is also crucial, as covered in Section 4.11.

The OSCOUT pin buffers the internal LO-signal output to the external PLL. This output should be ac-coupled and terminated at the far end (typically, the input to a prescaler) with a 50-Ω load. If a larger output level is desired, use a resistive termination of as much as 100 Ω. When a controlled-impedance PC board is used, the OSCOUT trace should match the termination impedance.

4.5 Mode Selection

The MAX2511 features four power-supply modes to preserve battery life. These modes are selected by control signals at the RXEN and TXEN pins, as follows:

RXEN	TXEN	MODE
Low	Low	Shutdown
Low	High	Transmit
High	Low	Receive
High	High	Standby

In the Shutdown mode, all part functions are off. In Standby, the LO and LO buffer are active. This allows an external PLL to remain up and running, thus avoiding any delay resulting from PLL settling. Transmit (Tx) mode enables the LO circuit, upconverter-mixer, transmitter VGA, and output driver-amplifier. Receive (Rx) mode enables the LO circuit, downconverter-mixer, limiting amplifier, and adjustable output-level amplifier.

4.6 200- to 440-MHz RF Applications

The MAX2511 can be used in applications where the 200-MHz to 440-MHz signal is an RF (rather than an IF) signal. In this case, Maxim recommends preceding the MAX2511 receiver section with a LNA that can operate over the same supply-voltage range. The MAX2630/MAX2633 family of amplifiers meets this requirement. However, because these amplifiers have single-ended inputs and outputs, it is necessary to ac-terminate the unused MAX2511 input ($\overline{\text{RXIN}}$) to ground with 47 nF.

4.7 Oscillator Tank Calculations

The on-chip oscillator requires a parallel-resonant tank circuit connected across TANK and $\overline{\text{TANK}}$. Figure 4.5 shows an example of such a circuit. Inductor L1 is resonated with the effective total capacitance of C_1 in parallel with the series combination of C_2, C_3, and $C_{D1/2}$. C_{D1} is the capacitance of one of the varactor diodes. Typically, $C_2 = C_3$ to maintain symmetery. The effective parasitic capacitance C_p (including the PC-board parasitics) is about 3.5 pF. The total or effective capacitance is given by the following equation:

$$C_{\text{EFF}} = \frac{1}{\frac{2}{C_2} + \frac{2}{C_{D1}}} + C_1 + C_p$$

Using this value for the tank circuit, the oscillator frequency is:

$$F_{\text{OSC}} = \frac{1}{6.28 \sqrt{L_1 C_{\text{EFF}}}}$$

The recommended values for L1 are:

fLO (MHz)	L1 (µH)
200 to 300	18
300 to 400	12
400 to 500	8.2

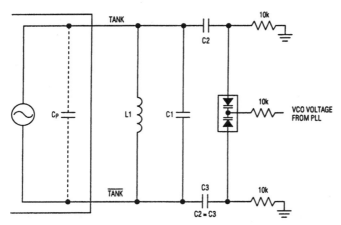

Figure 4.5 Oscillator tank schematic (*Maxim,* New Releases Data Book, *1998, p. 10-46*).

Starting with these L1 values, choose the remaining components according to the LO frequency range. Figure 4.3 shows some typical values for an LO frequency range of 400 MHz to 500 MHz.

For optimum performance (low phase noise) keep the tank Q as high as possible. For most of the MAX2511 applications (such as a first IF to second IF transceiver), the LO tuning range can be quite small because the IF frequencies are not tuned for channel selection. This allows a narrowband oscillator tank to be used, and provides better phase-noise and stability performance (than a wideband tank). The tank PC-board layout is also crucial, as covered in Section 4.11.

If the LO is to be driven (overdriven) from an external source, use the circuit shown in Fig. 4.6.

4.8 Impedance Matching for the Receiver Input

The receiver input pins RXIN and $\overline{\text{RXIN}}$ typically need an impedance-matching network, such as shown in Fig. 4.3. A shunt resistor across RXIN and $\overline{\text{RXIN}}$ can be used to set the terminating impedance. However, such a resistor causes a slight degradation of the noise figure.

The component values used in the matching network depend on the desired operating frequency, as well as the filter impedance. The RXIN and $\overline{\text{RXIN}}$ differential input impedances (in both series and parallel form) are:

Frequency (MHz)	Series impedance (ohms)	Equivalent R (ohms)	Parallel impedance C (pF)
100	274−j226	460	2.85
200	131−j186	395	2.86
300	79−j138	320	2.9
400	58−j105	248	2.9
500	48−82	188	2.9
600	43−62	132	2.9

4.9 Filter Sharing

Figure 4.7 shows how the external transmitter and receiver filters can be combined to minimize component count. This is particularly useful in half-duplex applications. The 10.7-MHz filter that is usually connected to the TXIN and $\overline{\text{TXIN}}$ pins can be the same filter that is connected at the LIMOUT and $\overline{\text{LIMOUT}}$. To use the same filter, connect TXIN to LIMOUT and $\overline{\text{TXIN}}$ to $\overline{\text{LIMOUT}}$.

The 425-MHz SAW filter needed at the RXIN and $\overline{\text{RXIN}}$ pins and the filter needed at the TXOUT and $\overline{\text{TXOUT}}$ pins can be shared in a similar manner. The RXIN and $\overline{\text{RXIN}}$ pins must be dc-blocked to prevent the bias voltage needed by the TXOUT and $\overline{\text{TXOUT}}$ pins from entering the receiver.

When sharing filters in this manner, matching networks for the transmitter output (TXOUT/$\overline{\text{TXOUT}}$), and receiver input (RXIN/$\overline{\text{RXIN}}$) must be modified. The receiver port input impedance must be the parallel combination of the transmitter and receiver ports

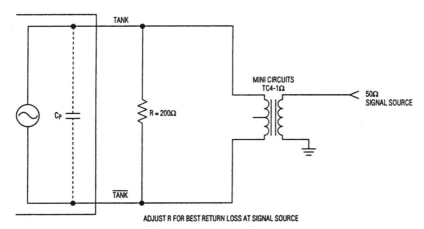

Figure 4.6 Overdriving the on-chip oscillator (*Maxim*, New Releases Data Book, 1998, p. 10-46).

in the Rx mode. In this case, the receiver port is active, but the transmitter port adds an additional parasitic impedance. (The MAX2511 data sheet shows this characteristic in graph form.)

In the Transmit mode, the RXIN and $\overline{\text{RXIN}}$ inputs provide back termination for the TXOUT and $\overline{\text{TXOUT}}$ outputs so that a single IF filter can be connected, as shown in Fig. 4.7. Using this technique, the matching network can be adjusted so that the input VSWR can be less than 1.5:1 in the Rx mode and the output VSWR is less than 2:1 in the Tx mode.

4.10 Receiver IF Filter

The interstage 10.7-MHz filter, located between the MIXOUT and LIMIN pins (Figs. 4.1 and 4.7), is not shared. This filter prevents the limiter from acting on any undesired signals that are present at the mixer output (such as LO feedthrough, out-of-band channel leakage, and other mixer products). The filter is also set up to pass dc-bias voltage from the V_{REF} pin into the LIMIN and MIXOUT pins through two 330-Ω filter-termination resistors. If the filter can provide a dc-shunt path, such as a transformer/capacitor-based filter or some LC filters, the two resistors can be combined into one parallel-equivalent 165-Ω resistor, as shown in the inset of Fig. 4.7. This reduces the component count by one resistor.

4.11 Optimizing MAX2511 Layout

To minimize coupling between different sections of the chip, the ideal power-supply layout is (as always) the star configuration, which has a heavily decoupled central V_{CC} node. The V_{CC} traces branch out from the central node, with each going to one V_{CC} node on the MAX2511. At the end of each of these traces is a bypass capacitor that is good at the RF frequency of interest.

The star configuration provides local decoupling at each V_{CC} pin. At high frequency, any signal leaking from a supply pin sees a relatively high impedance (formed by the V_{CC} trace impedance) to the central V_{CC} node, and an even higher impedance to any other supply pin.

Place the V_{REF} decoupling capacitor (0.1 μF, typical) as close to the MAX2511 as possible for best interstage filter performance. Use a high-quality, low-ESR (equivalent series resistance) capacitor for best results.

The TXOUT and $\overline{\text{TXOUT}}$ ports require a bias network that consists of two inductors to V_{CC} (for differential drive) and (optionally) a back-termination resistor for matching to an external filter. The RXIN and $\overline{\text{RXIN}}$ ports also need an impedance-matching network. Both networks

IF Transceiver with Limiter and RSSI

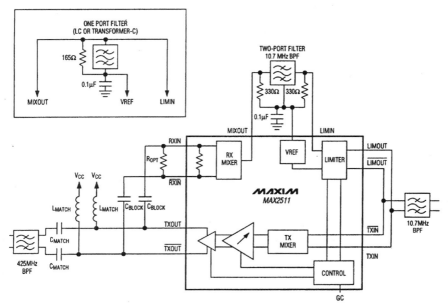

Figure 4.7 Filter sharing (*Maxim,* New Releases Data Book, *1998, p. 10-47*).

should be symmetrical and as close to the chip as possible. See Fig. 4.3 for typical connections. If the PC board has a ground plane (generally recommended), cut out the ground plane under the matching-network components to reduce parasitic capacitance.

For this (and most similar ICs), the oscillator-tank circuit layout is crucial. Parasitic PC-board capacitance, as well as trace inductance, can affect oscillation frequency. Keep the tank layout symmetrical, tightly packed, and as close to the IC as possible. Again, if a ground-plane PC board is used, the ground plane should be cut out under the oscillator components to reduce parasitic capacitance.

Chapter 5

RF Power Transistors for 900 MHz

This chapter is devoted to transistors suitable for operation in 900-MHz RF circuits. The MAX2601 and MAX2602 are selected as examples. The characteristics of these transistors are optimized for use in portable cellular and wireless equipment that operates from three NiCd/NiMH cells or one Li-ion cell.

5.1 Transistor Characteristics

Figures 5.1, 5.2, and 5.3 show the pin configurations, pin descriptions, and typical connections (test circuit), respectively. These transistors deliver 1 W of RF power from a 3.6-V supply with efficiency of 58% when biased for constant-envelope applications (such as FM or FSK, frequency-shift keying).

The transistors are high-performance silicon bipolar devices in power-enhanced, 8-pin SO packages. The base and collector connections use two pins each to reduce series inductance. The emitter connects to three (MAX2602) or four (MAX2601) pins in addition to a back-side heat slug, which solders directly to a PC-board ground to reduce emitter inductance and improve thermal dissipation. The transistors are intended for use in the common-emitter configuration for maximum power gain and power-added efficiency.

The transistors can be used as the final stage in a discrete or modular power amplifier. Silicon bipolar technology eliminates the need for voltage inverters and sequencing circuitry (as required by GaAsFET power amplifiers). Also, a drain switch is not required to turn off the transistors. This increases operating time in two ways: by allowing lower system and end-of-life battery voltage, and by eliminating the wasted power from a drain-switch device.

TOP VIEW

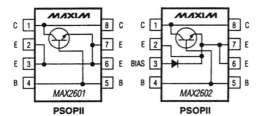

Figure 5.1 MAX2601/2602 pin configurations (*Maxim*, New Releases Data Book, *1998*, p. 10-49).

PIN		NAME	FUNCTION
MAX2601	MAX2602		
1, 8	1, 8	C	Transistor Collector
2, 3, 6, 7, Slug	2, 6, 7, Slug	E	Transistor Emitter
—	3	BIAS	Anode of the Biasing Diode that matches the thermal and process characteristics of the power transistor. Requires a high-RF-impedance, low-DC-impedance (e.g., inductor) connection to the transistor base (Pin 4). Current through the biasing diode (into Pin 3) is proportional to 1/15 the collector current in the transistor.
4, 5	4, 5	B	Transistor Base

Figure 5.2 MAX2601/2602 pin descriptions (*Maxim*, New Releases Data Book, *1998*, p. 10-51).

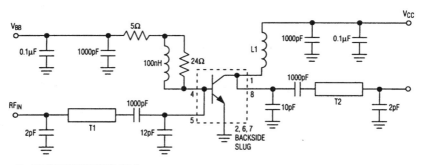

L1 = COILCRAFT A05T INDUCTOR, 18.5nH
T1, T2 = 1", 50Ω TRANSMISSION LINE ON FR-4

Figure 5.3 MAX2601/2602 typical connections for operation at 836 MHz (*Maxim*, New Releases Data Book, *1998*, p. 10-52).

5.2 Current-Mirror Bias

As shown in Fig. 5.1, the MAX2602 includes a high-performance silicon bipolar RF-power transistor and a thermally matched biasing diode. Figure 5.4 shows how the diode is used to create a bias network that accurately controls the power-transistor collector current when the temperature changes (the diode characteristics match those of the transistor).

The biasing diode is a scaled version of the transistor base-emitter junction. The scale factor is such that the diode current is $\frac{1}{15}$ of the transistor quiescent current. As shown in the typical circuit of Fig. 5.4, diode Q1 is provided with a constant current through R_{BIAS}. If the temperature increases, the current through both Q1 and Q2 tends to increase. In turn, this causes an increase in the voltage drop across R_{BIAS}, thus lowering the voltage on the base of Q2 (to offset the initial current increase). As a result, the Q1 collector current remains constant, in spite of temperature variations.

In most cases, the diode can simply be connected to the supply through a fixed resistor, with the diode-resistor junction connected to the transistor base. However, for optimum performance when large supply variations are anticipated, connect the diode-resistor junction to the transistor through an inductor, such as an RFC (radio frequency choke), as shown in Fig. 5.4. Also, it might be necessary to decouple the diode-resistor junction to ground through a surface-mounted chip capacitor (with a value of at least 1000 pF).

5.3 Optimum Port Impedance

As is the case with any transistor, the source and load terminating impedances (Z_S and Z_L) presented to the base and collector must be of the correct value to ensure optimum performance. Z_S and Z_L have a direct impact on transistor gain, output power, and linearity. When the MAX2601/2602 are used as power transistors, simply applying the conjugate of the transistor input and output impedances (calculated, as described in Chapter 2, or by computer calculation using S-parameters, y-parameters, etc.) will not produce optimum performance. Instead, special Z_S and Z_L values must be used.

For maximum efficiency at $V_{BB} = 0.75$ V and $V_{CC} = 3.6$ V, the optimum power-transistor source and load impedances are:

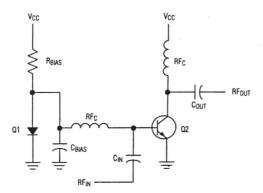

Figure 5.4 MAX2602 current-mirror bias circuit (*Maxim*, New Releases Data Book, *1998*, p. 10-52).

At 836 MHz: $Z_S = 5.5 + j2.0$
$Z_L = 6.5 + j1.5$
At 433 MHz: $Z_S = 9.5 - j2.5$
$Z_L = 8.5 - j1.5$

These impedance values arise because parasitic impedances (mostly inductive) are at the input and output pins, as shown in Fig. 5.5. These internal bond and package inductances should be included as part of the end-application matching networks.

5.4 Optimizing MAX2601/2602 Layout

The most important connection to the transistors is at the back side. As shown in Fig. 5.3, the back-side slug should connect directly to the PC-board ground plane if the plane is on the top side. When the plane is on the bottom or is buried, the slug must be connected through many plated through-holes.

For maximum gain, the slug connection should have very little self-inductance. Also, because the slug connection provides the thermal path for heat dissipation, the connection must have low thermal impedance, and the ground plane should be large.

Notice that for the MAX2601, pins 2, 3, 6, and 7 are grounded. For the MAX2602, pin 3 is connected to the bias circuit, as shown in Fig. 5.4.

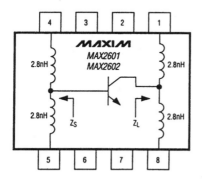

Figure 5.5 Parasitic inductances (*Maxim*, New Releases Data Book, *1998, p. 10-53*).

Chapter 6

IC Oscillator for 650 to 1050 MHz

This chapter is devoted to IC oscillators. The MAX2620 is selected as an example. This IC combines a low-noise oscillator with two output buffers in a low-cost, plastic surface-mount, ultra-small µMAX package. As a result, the IC integrates functions that are typically achieved with discrete components.

6.1   MAX2620 Characteristics

Figures 6.1, 6.2, and 6.3 show a typical operating circuit, pin descriptions, and a test circuit, respectively. The oscillator shows low phase noise when properly terminated with an external varactor-tuned resonant tank. The two buffered outputs provide for driving mixers or prescalers. The buffers also provide load isolation to the oscillator and prevent pulling because of load-impedance changes (often a problem with discrete oscillators, particularly at high frequencies).

Power consumption is typically 27 mW in the Operating mode (V_{CC} = 3.0 V) and drops to less than 0.3 µW in Standby mode. The IC operates from a single +2.7-V to +5.25-V supply.

6.2   Oscillator Circuit

The oscillator is a common-collector, negative-resistance type that uses the IC internal parasitic elements to create a negative resistance at the base-emitter port. The transistor oscillator is optimized for low-noise operation. Base and emitter leads are provided as external connections for a feedback capacitor and resonator.

A resonant circuit, tuned to the appropriate frequency and connected to the base lead, will cause oscillation. [As shown by Fig. 6.1, a ceramic res-

112 Chapter Six

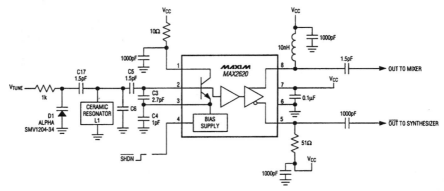

Figure 6.1 MAX2620 typical operating circuit (*Maxim*, New Releases Data Book, *1998*, p. 10-55).

PIN	NAME	FUNCTION
1	$V_{CC}1$	Oscillator DC Supply Voltage. Decouple $V_{CC}1$ with 1000pF capacitor to ground. Use a capacitor with low series inductance (size 0805 or smaller). Further power-supply decoupling can be achieved by adding a 10Ω resistor in series from $V_{CC}1$ to the supply. Proper power-supply decoupling is critical to the low noise and spurious performance of any oscillator.
2	TANK	Oscillator Tank Circuit Connection. Refer to the *Applications Information* section.
3	FDBK	Oscillator Feedback Circuit Connection. Connecting capacitors of the appropriate value between FDBK and TANK and between FDBK and GND tunes the oscillator's reflection gain (negative resistance) to peak at the desired oscillation frequency. Refer to the *Applications Information* section.
4	$\overline{SHDN}$	Logic-Controlled Input. A low level turns off the entire circuitry such that the IC will draw only leakage current at its supply pins. This is a high-impedance input.
5	$\overline{OUT}$	Open-Collector Output Buffer (complement). Requires external pull-up to the voltage supply. Pull-up can be resistor, choke, or inductor (which is part of a matching network). The matching-circuit approach provides the highest-power output and greatest efficiency. Refer to Table 1 and the *Applications Information* section. $\overline{OUT}$ may be used with OUT in a differential output configuration.
6	GND	Ground Connection. Provide a low-inductance connection to the circuit ground plane.
7	$V_{CC}2$	Output Buffer DC Supply Voltage. Decouple $V_{CC}2$ with a 1000pF capacitor to ground. Use a capacitor with low series inductance (size 0805 or smaller).
8	OUT	Open-Collector Output Buffer. Requires external pull-up to the voltage supply. Pull-up can be resistor, choke, or inductor (which is part of a matching network). The matching-circuit approach provides the highest-power output and greatest efficiency. Refer to Table 1 and the *Applications Information* section. OUT may be used with $\overline{OUT}$ in a differential output configuration.

Figure 6.2 MAX2620 pin descriptions (*Maxim*, New Releases Data Book, *1998, p. 10-59*).

onator (L1) can serve to set the operating frequency.] Varactor diodes can be used in the resonant circuit to create a VCO, if desired. The oscillator is internally biased to an optimal operating point and the base/emitter leads must be capacitively coupled because of the bias voltages present.

6.3 Output Buffers

The output buffers OUT and $\overline{OUT}$ form an open-collector differential pair configuration and provide load isolation to the oscillator. The outputs can be used differentially to drive an IC mixer. In an alternate form, isolation is provided between the buffer outputs when one output drives a mixer (either upconverter or downconverter) and the other output drives a prescaler. The isolation provided by this configuration prevents prescaler noise from corrupting the oscillator signal. A logic-controlled $\overline{SHDN}$ pin turns off all bias to the IC when it is pulled low. This stops the oscillator and disables the output buffers.

6.4 Oscillator Tank-Circuit Design

Figure 6.4 shows a one-port model of the oscillator tank circuit. Figure 6.5 shows a Smith-chart 1/S11 representation at the test port (Fig. 6.3). The 1/S11 representation is part of a classic RF design technique involving S (scattering) parameters. In this case, S11 is used because it maps inside the unit circle of the Smith chart when the oscillator device shows negative resistance.

The oscillator can be tuned over a given range by the varactor diode (Fig. 6.1) as long as the oscillator shows negative resistance over the range. This is done by making the resonant tank have a positive real part (R) that is one third to one half of the magnitude of the negative real part of the oscillator device ($-R$), as well as a reactive part (jX)

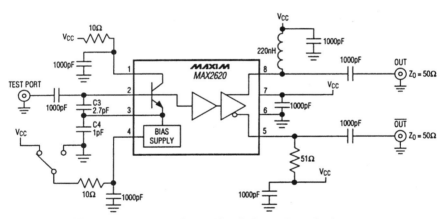

Figure 6.3 MAX2620 test circuit (*Maxim*, New Releases Data Book, *1998, p. 10-60*).

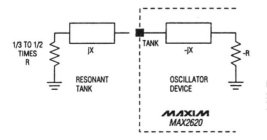

Figure 6.4 Oscillator tank circuit model (*Maxim*, New Releases Data Book, *1998, p. 10-61*).

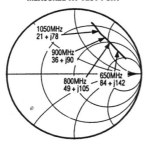

Figure 6.5 MAX2620 Smith-chart 1/S11 representation (*Maxim*, New Releases Data Book, *1998, p. 10-58*).

that is opposite in sign to the reactive component of the oscillator device ($-jX$), as shown in Fig. 6.4.

Keeping the resonant-tank real part between one third and one half of the magnitude of the device's negative real part ensures that oscillations will start. (Oscillator start-up problems are covered in Section 2.16.) After start up, the oscillator negative resistance decreases and becomes equal to the real part (the circuit losses) in the resonant tank circuit.

The negative-resistance characteristics can be tuned by selecting the values of the feedback capacitor connected between tank and FDBK (C3) and from FDBK to ground (C4). (Feedback-capacitor ratios for oscillators are also covered in Section 2.16.) To get the approximate values of C_3 and C_4 for a given frequency, use the following: $R_{TANK} = gmX_{C3}X_{C4}$, where $gm = 0.018$ and X_{C3} and $X_{C4} = $ the reactance of C3 and C4 at the desired frequency. To get a more precise value, make a one-port measurement into the TANK (at the test port in Fig. 6.3) using a vector network analyzer.

For the real world, use the values shown in Figs. 6.1 and 6.3 ($C_3 = 2.7$ pF and $C_4 = 1$ pF) as a starting point for design. These values provide for oscillation at 900 MHz when L1 is a Trans-Tech S8800LPQ1357B ceramic resonator and C_6 is 1 pF, or when L1 is a 5-nH high-Q inductor and C_6 is 1.5 pF. The ceramic resonator provides lower phase noise. With either the inductor or resonator, keep C_5 and C_{17} (Fig. 6.1) as small as possible while still maintaining the desired frequency and tuning range.

6.5 Matching the Output Buffers

Both the OUT and $\overline{\text{OUT}}$ outputs are open collectors and must be pulled up to the supply by external components. Where maximum transfer of energy is not crucial, the outputs can simply be terminated in the characteristic impedance of the system (typically 50 Ω, using a standard-value 51-Ω resistor), as shown for $\overline{\text{OUT}}$ in Figs. 6.1 and 6.3.

For maximum transfer of energy, use a choke as the supply pull-up, such as shown for OUT. (The value of 220 nH is for a frequency of 900 MHz.) Using an inductor makes it possible to match the buffer output precisely to the system impedance at a given frequency. The following load impedances (R and X) are recommended for the MAX2620 frequency range:

Frequency (MHz)	Real part (R in ohms)	Imaginary part (X in ohms)
650	17.5	62.3
750	17.2	50.6
850	10.9	33.1
950	7.3	26.3
1050	6.5	22.7

Chapter 7

General-Purpose Amplifiers for VHF to Microwave

This chapter is devoted to typical IC amplifiers for use in wireless/RF circuits. The MAX2630 through 2633 are selected as examples. These ICs operate from a single +2.7-V to +5.5-V supply, and have a flat gain response to 900 MHz. Their low noise figure and low supply current make the ICs ideal for receiver, buffer, and transmitter IF applications, including cordless phones, cellular phones, TV tuners, set-top boxes, land-mobile radios, PCs (personal communicating systems), GPS (global positioning systems), WLAN (wireless local area networks), and WLL (wireless local loops).

7.1 MAX2630 through 2633 Characteristics

Figures 7.1, 7.2, 7.3, and 7.4 show a typical operating circuit, pin configurations, ratings and characteristics, and pin descriptions, respectively. The MAX2630/31 are biased internally, eliminating the need for external bias resistors or inductors. The MAX2632/33 have a user-selectable supply current, which can be adjusted by adding a single external resistor. This allows customized output power and gain, according to specific application requirements.

The MAX2631/33 feature a shutdown pin that allows the ICs to be powered down to less than 1 µA of supply current. Aside from a single bias resistor required for the MAX2632/33, the only external components needed for this family of IC amplifiers are input- and output-blocking capacitors and a V_{CC} bypass capacitor.

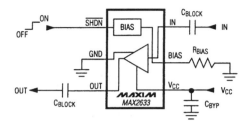

Figure 7.1 MAX2630 typical operating circuit (*Maxim,* New Releases Data Book, *1998, p. 10-63*).

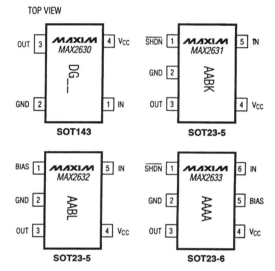

Figure 7.2 MAX2630 through MAX2633 pin configurations (*Maxim,* New Releases Data Book, *1998, p. 10-63*).

7.2 External Components for Typical Operating Circuits

Figures 7.5, 7.6, and 7.7 show typical operating circuits for the MAX2630, MAX2631, and MAX2632, respectively. In most cases, external capacitors for the input and output are required to block dc bias voltages generated by the amplifiers from interfering with adjacent circuitry. The blocking capacitors must be large enough to contribute negligible reactance in a 50-Ω system at the minimum operating frequency. Use the following to calculate the minimum blocking-capacitor value:

$$C_{\text{BLOCK}} \text{ (in pF)} = \frac{53{,}000}{f}$$

where f (in MHz) is the minimum operating frequency.

7.3 Optimizing MAX2630 through MAX2633 Layout

Figures 7.8, 7.9, 7.10, and 7.11 show examples of recommended PC-board layout for the MAX2630, MAX2631, MAX2632, and MAX2633, respectively. All layouts use FR-4 with a 31-mil layer thickness between the RF lines and the ground plane.

In all cases, the V_{CC} pin must be RF-bypassed for correct operation. This is done by connecting a capacitor between the V_{CC} pin and ground, as close to the IC package as is practical. Use the same equation given in Section 7.2 (for dc blocking-capacitor values) to calculate the bypass-capacitor value. If the PC board has long V_{CC} lines, additional bypassing might be required. This can be done farther away from the IC package (if needed).

ABSOLUTE MAXIMUM RATINGS

V_CC to GND ....-0.3V to 6V
Input Power..5dBm
OUT Current ....±12mA
IN to GND Voltage ....-1.2V to 1.2V
Bias to GND Voltage ....0.0V to 3V
Voltage at $\overline{SHDN}$ Input
 (MAX2631/MAX2633)..........................-0.3V to (V_{CC} + 0.3V)
Current into $\overline{SHDN}$ Input (MAX2631/MAX2633).................100µA

Continuous Power Dissipation (T_A = +70°C)
 SOT143 (derate 4mW/°C above +70°C).....................320mW
 SOT23-5 (derate 7.1mW/°C above +70°C)..................571mW
 SOT23-6 (derate 7.1mW/°C above +70°C)..................571mW
Operating Temperature Range ....-40°C to +85°C
Junction Temperature...+150°C
Storage Temperature Range ....-65°C to +150°C
Lead Temperature (soldering, 10sec)...........................+300°C

Stresses beyond those listed under "Absolute Maximum Ratings" may cause permanent damage to the device. These are stress ratings only, and functional operation of the device at these or any other conditions beyond those indicated in the operational sections of the specifications is not implied. Exposure to absolute maximum rating conditions for extended periods may affect device reliability.

ELECTRICAL CHARACTERISTICS

(V_{CC} = +3V, Z_0 = 50Ω, f_{IN} = 900MHz, R_{BIAS} = 10kΩ (MAX2632/MAX2633), $V\overline{SHDN}$ = V_{CC} (MAX2631/MAX2633), T_A = +25°C, unless otherwise noted.)

PARAMETERS	CONDITIONS		MIN	TYP	MAX	UNITS
Operating Temperature Range	(Note 1)		-40		85	degrees
Supply Voltage			2.7		5.5	V
Power Gain	T_A = +25°C		11	13.4	16.5	dB
	T_A = T_{MIN} to T_{MAX} (Note 1)		9.4		18.4	
Noise Figure				3.8		dB
Output 1dB Compression Point				-11		dBm
Output IP3				-1		dBm
Input Voltage Standing-Wave Ratio	f_{IN} = 800MHz to 1000MHz			1.3:1		
Output Voltage Standing-Wave Ratio	f_{IN} = 800MHz to 1000MHz			1.25:1		
Supply Current	R_{BIAS} = 40kΩ			1.3	1.5	mA
	R_{BIAS} = 10kΩ	V_{CC} = 3V, T_A = +25°C	5.5	6.5	8.0	
		V_{CC} = 3V, T_A = T_{MIN} to T_{MAX} (Note1)	4.2	6.5	9.2	
		V_{CC} = 2.7V to 5.5V, T_A = +25°C	5.2	6.5	11.0	
	R_{BIAS} = 500Ω		15	17		
Shutdown Supply Current	MAX2631/MAX2633			<0.1	1	µA
$\overline{SHDN}$ Input Low Voltage	MAX2631/MAX2633, V_{CC} = 2.7V to 5.5V				0.45	V
$\overline{SHDN}$ Input High Voltage	MAX2631/MAX2633, V_{CC} = 2.7V to 5.5V		2.0			V
$\overline{SHDN}$ Input Bias Current	MAX2631/ MAX2633	$V\overline{SHDN}$ = V_{CC}			30	µA
		$V\overline{SHDN}$ = GND			1	

Note 1: Guaranteed by design and characterization.

Figure 7.3 MAX2630 through MAX2633 ratings and characteristics (*Maxim*, New Releases Data Book, *1998*, p. 10-64).

PIN				NAME	FUNCTION
MAX2630	MAX2631	MAX2632	MAX2633		
1	5	5	6	IN	Amplifier Input. Use a series blocking capacitor with less than 3Ω reactance at your lowest operating frequency.
2	2	2	2	GND	Ground Connection. For optimum performance, provide a low-inductance connection to the ground plane.
3	3	3	3	OUT	Amplifier Output. Use a series blocking capacitor with less than 3Ω reactance at your lowest operating frequency.
4	4	4	4	V_{CC}	Supply Connection. Bypass directly at the supply pin. The value of the bypass capacitor is determined by the lowest operating frequency, and is typically the same as the blocking capacitor value. Additional bypassing may be necessary for long V_{CC} lines.
—	1	—	1	$\overline{SHDN}$	Shutdown Input. Driving $\overline{SHDN}$ with a logic low turns off the amplifier.
—	—	1	5	BIAS	Bias Resistor Connection. Connect a resistor to GND to set the bias current. See the Supply Current vs. R_{BIAS} graph in the *Typical Operating Characteristics*.

Figure 7.4 MAX2630 through MAX2633 pin descriptions (*Maxim*, New Releases Data Book, *1998, p. 10-66*).

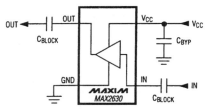

Figure 7.5 MAX2630 typical operating circuit (*Maxim*, New Releases Data Book, *1998, p. 10-69*).

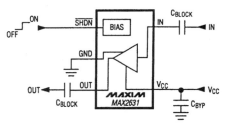

Figure 7.6 MAX2631 typical operating circuit (*Maxim*, New Releases Data Book, *1998, p. 10-69*).

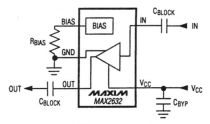

Figure 7.7 MAX2632 typical operating circuit (*Maxim*, New Releases Data Book, *1998, p. 10-69*).

General-Purpose Amplifiers for VHF to Microwave   121

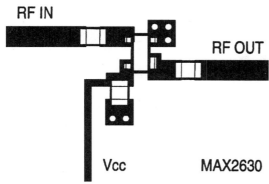

Figure 7.8 MAX2630 recommended PC-board layout (*Maxim*, New Releases Data Book, *1998*, p. 10-70).

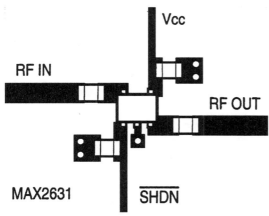

Figure 7.9 MAX2631 recommended PC-board layout (*Maxim*, New Releases Data Book, *1998*, p. 10-70).

Proper grounding of the GND pin is essential. If the PC board uses a top-side RF ground, connect this ground directly to the GND pin. For boards where the ground plane is not on the component side, the best technique is to connect the GND pin to the plane with a plated through-hole close to the IC package.

The $\overline{\text{SHND}}$ pin (MAX2631/33) does not usually require bypassing, except in very noisy applications. When RF filtering is needed, use a bypass capacitor similar to the one used on V_{CC}. Because negligible current flows into the $\overline{\text{SHND}}$ pin, additional RF filtering can be performed with a series resistor.

To set the MAX2632/33 supply current, connect a resistor from the BIAS pin to ground (Fig. 7.7). To estimate the value for the bias resistor, use the graph of Fig. 7.12.

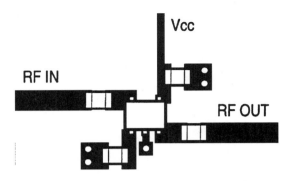

MAX2632

Figure 7.10 MAX2632 recommended PC-board layout (*Maxim,* New Releases Data Book, *1998, p. 10-70*).

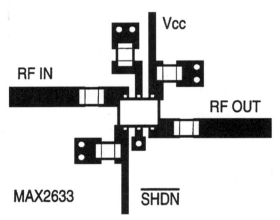

Figure 7.11 MAX2633 recommended PC-board layout (*Maxim,* New Releases Data Book, *1998, p. 10-70*).

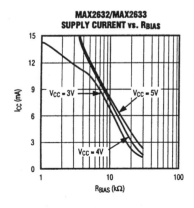

Figure 7.12 Graph for selecting bias resistance (*Maxim,* New Releases Data Book, *1998, p. 10-66*).

Chapter 8

Applications for Special-Purpose RF ICs

This chapter is devoted to optimizing special-purpose RF ICs. A cross-section of such ICs is covered, with specific applications for each IC. The circuits here can be used immediately the way they are or, by altering component values, as a basis for the design of similar circuits. All of the general optimizing and design information in Chapters 1 and 2 applies to the examples in this chapter. However, each IC has special requirements, all of which are covered in detail.

8.1 Wideband VHF Antenna Booster

This section describes a 40- to 250-MHz antenna amplifier module that is suitable for any 50-Ω VHF system. The module is constructed on a single-sided PC board. Layout is not crucial, as long as the usual precautions for VHF construction are observed.

8.1.1 Circuit and characteristics (wideband VHF)

Figures 8.1 and 8.2 show the circuit and performance characteristics, respectively. The GEC Plessey SL560 is used as the IC amplifier. An unusual feature of the module is a transmission-line transformer (T1) that provides "noiseless" emitter feedback to the SL560.

8.1.2 Optimizing wideband VHF booster layout

The following points should be observed when constructing the module:

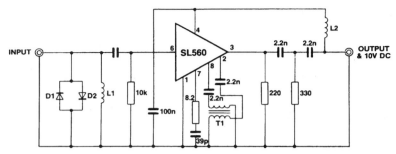

Figure 8.1 Wideband VHF antenna booster (*GEC Plessey Semiconductors, Professional Products IC Handbook, p. 4-3*).

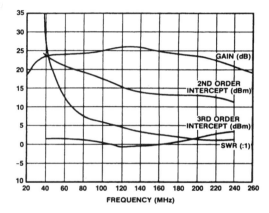

Figure 8.2 Performance characteristics for antenna booster (*GEC Plessey Semiconductors*, Professional Products IC Handbook, *p. 4-3*).

- Diodes D1 and D2 are any general-purpose silicon devices.
- L1 is 8 turns of #26 wire, $\frac{1}{8}$-inch internal diameter.
- L2 is 20 turns of #26 wire, $\frac{3}{16}$-inch internal diameter.

T1 consists of two lengths of #34 wire that is approximately six inches in length, twisted together (eight twists per inch) and wound on a six-hole ferrite bead (Mullard FX1898, or equivalent).

Resistor and capacitor values are not crucial, but should be close to the values shown for wideband response.

The module is powered through the coax cable, and requires 10 V at 30 mA.

8.2 Tuned Amplifier (Matching for Increased Gain)

This section describes a 100-MHz tuned amplifier with a 35-dB power gain, suitable for a 50-Ω system. The IC involved, a GEC Plessey SL6140 wideband AGC amplifier, provides only a 15-dB gain when untuned. By tuning (matching) the inputs and outputs of the IC, the gain is increased substantially. However, the tuning function decreases bandwidth. As covered in Chapters 1 and 2, the overall amplifier bandwidth depends on the Q factor of the tuning/matching circuits used.

8.2.1 Circuit description (tuned amplifier)

Figure 8.3 shows the circuit for a single-ended amplifier with tuned input and output networks. Figure 8.4 shows the circuit for a differential-tuned output.

The input circuit consists of a parallel LC network connected across the differential inputs of the SL6140. The input signal is applied to one input through coupling capacitor C1. The other input is decoupled. Capacitor C1 also forms part of the impedance-matching network that matches the 50-Ω source with the high-impedance input of the SL6140. Figure 8.5 shows a Smith chart presentation of the SL6140 input, normalized to 50 Ω.

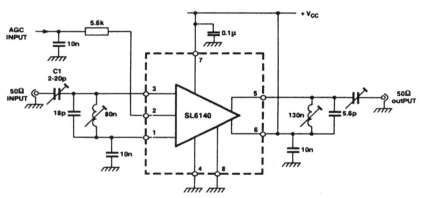

Figure 8.3 Single-ended tuned amplifier (*GEC Plessey Semiconductors*, Professional Products IC Handbook, *p. 4-5*).

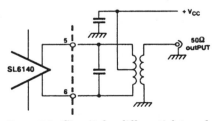

Figure 8.4 Circuit for differential tuned output (*GEC Plessey Semiconductors*, Professional Products IC Handbook, *p. 4-5*).

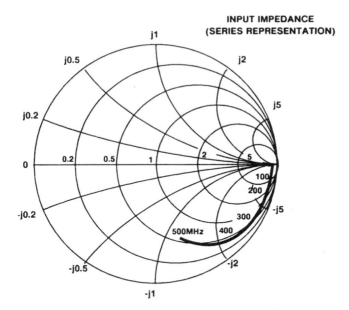

Figure 8.5 Smith chart presentation of SL6140 input (*GEC Plessey Semiconductors,* Professional Products IC Handbook, p. 4-6).

The tuned frequency is given by the following:

$$f = \frac{1}{6.28 \frac{\sqrt{LCC_1}}{\sqrt{C + C_1}}}$$

The output circuit consists of a parallel LC network connected from one of the open-collector outputs of the device to V_{CC}. The other output is connected directly to V_{CC}. The coupling capacitor (C2) and LC network transform the 50-Ω load to a high-impedance load for the open-collector output of the IC.

8.2.2 Optimizing for maximum gain

The gain can be optimized by adjusting C1 and C2 (the unmarked trimmer capacitor at the 50-Ω output). However, if too high an impedance is "seen" by the input or output of the IC, the circuit might oscillate. Notice that C1 and C2 set the bandwidth and gain, but L_1 and L_2 set the tuned frequency.

As is typical for most RF amplifier circuits, a tradeoff occurs between bandwidth and gain. As a practical optimizing technique, adjust for maximum gain at the minimum bandwidth required by the particular application.

8.2.3 Maximum gain with differential output

The differential-tuned output circuit shown in Fig. 8.4 provides about 6 dB of additional gain over that provided by the single-ended circuit of Fig. 8.3. However, the Fig. 8.4 circuit requires a transformer. The primary winding is connected across the outputs of the IC, with V_{CC} being provided by a center tap. The capacitor in parallel with the transformer primary is selected to provide resonance at the center of the desired operating frequency. The transformer secondary must match the 50-Ω system.

8.2.4 Optimizing the tuned amplifier layout

A ground plane with a 50-Ω track from the matching networks to the 50-Ω source and load should be used. The matching networks and decoupling capacitors should be positioned as close to the device as is practical. If very high gain (with low bandwidth) is required, it might be necessary to provide shielding between the input and output to prevent oscillation.

8.3 High-Performance Mixer

This section describes a mixer circuit using a single IC with few external components. The IC involved is a GEC Plessey SL6440 that provides a mixer function far superior to that provided by classic mixer circuits (diode, FET, diode ring, and quad FETs).

Whenever signals are mixing, such as the mixing of the local-oscillator (LO) and RF signals in a receiver, certain undesirable effects occur. The two most common problems are intermodulation distortion (IMD) and phase noise. Both of these affect the dynamic range (DR) of the receiver. Before getting into the IC mixer, review the problems commonly associated with mixer circuits.

8.3.1 IMD, phase noise, and dynamic range

Figure 8.6 shows the spectrum-analyzer display of IMD. The effects of intermodulation are similar to those produced by mixing and harmonic production (see Fig. 2.21). That is, when two signals (f_1 and f_2) are mixed, outputs of $2f_2 - f_1$, $2f_1 - f_2$, $2f_1$, $2f_2$, etc., are produced. The levels of these signals depend on the transfer function of the mixing device (diode, FET, IC, etc.).

The effects of intermodulation are to produce unwanted signals that degrade the effective signal-to-noise (S/N) ratio of the wanted signal. In the case of radio communications, the receiver must be able to pick out a very weak wanted signal from a background of noise; at the same time, the receiver rejects a large number of stronger unwanted signals. With an increase in the number of input signals, the effects of intermodulation are multiplied, and cause a rise in the noise floor of the receiver.

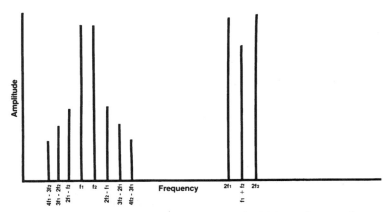

Figure 8.6 Spectrum-analyzer display of IMD (*GEC Plessey Semiconductors, Professional Products IC Handbook, p. 4-25*).

The third-order intermodulation products ($2f_1 - f_2$, $2f_1 - f_3$, $2f_2 - f_1$, $2f_2 - f_3$, $2f_3 - f_1$, and $2f_3 - f_2$) are of most concern. The amplitude of such products is proportional to the cube of the input signal level. That is, an increase of 3 dB in input level produces an increase of 9 dB in the levels of the intermodulation products.

Figure 8.7 shows the amplitude relationship of the fundamental and intermodulation-produced signals, and indicates where the two amplitudes intercept (the third-order intercept point, sometimes known as *IM3*). Actually, the third-order intercept point is a theoretical concept. The two amplitudes never intercept because of gain compression in a practical receiver circuit. However, the intermodulation-produced noise floor in a receiver is related to the theoretical intercept point, and both affect the dynamic range.

Figure 8.8 shows the concept of dynamic range when discussing intermodulation. An intermodulation product below the receiver noise floor can be ignored. The usable dynamic range is that input range between the noise floor and the input level at which the intermodulation product reaches the noise floor. This can be expressed as:

$$DR = \frac{2}{3}(I_3 - NF)$$

where DR is the dynamic range in dB, I_3 is the intermodulation input intercept point in dBm, and NF is the noise floor in dBm.

Figure 8.9 shows the spectrum-analyzer presentation for the signal-mixing process in a superhet receiver, where the incoming signal mixes with the LO to produce the intermediate frequency (IF). Figure 8.10 shows the effect of phase noise on the LO, where the noise sidebands

Applications for Special-Purpose RF ICs   129

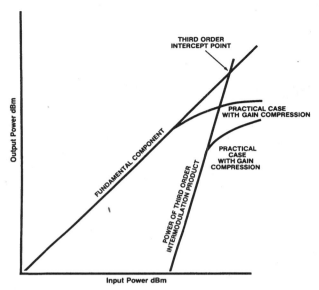

Figure 8.7 Amplitude relationship of fundamental and IMD, showing third-order intercept (*GEC Plessey Semiconductors, Professional Products IC Handbook, p. 4-26*).

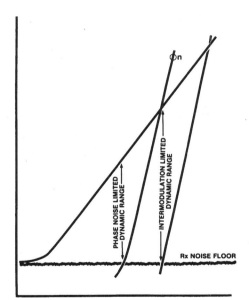

Figure 8.8 Concept of intermodulation and dynamic range (*GEC Plessey Semiconductors,* Professional Products IC Handbook, *p. 4-27*).

of the LO mix with a strong off-channel signal to produce the IF. (The condition shown in Fig. 8.10 is called *reciprocal mixing* in some literature.) In practical terms, this means that the phase noise performance of the LO affects the capability of the receiver to reject off-channel signals, and thus determines receiver selectivity (along with other factors outside of the mixer circuit).

Now that you know the basic problems in mixers, see how the SL6440 IC performs in relation to other classic mixer circuits.

8.3.2 Characteristics of mixer types

The following summarizes the advantages and disadvantages of various mixer types.

A *single diode mixer* has a wide bandwidth, requires little power from the LO, and is generally the lowest in cost. However there is no rejection of any signals, including the signals that result from mixing. No isolation is between input and output, and the input impedance is generally very low.

A *single FET* has very low noise, no signal loss (possibly some signal gain), and the cost is only slightly higher than that of the single diode. However, the IMD produced by a single FET can be quite severe. If the FET was a true square-law device, it would have no third-order products, but this is not practical. Again, there is no isolation, AM (amplitude

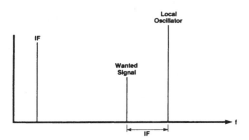

Figure 8.9 Spectrum-analyzer presentation for superhet signal mixing (*GEC Plessey Semiconductors*, Professional Products IC Handbook, *p. 4-29*).

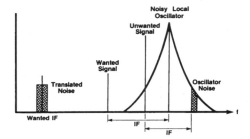

Figure 8.10 Reciprocal mixing (*GEC Plessey Semiconductors*, Professional Products IC Handbook, *p. 4-29*).

modulation) rejection is poor, and a single FET is easily overloaded and has low input impedance.

A *diode ring* has better IMD characteristics than a single diode or FET because the distortion produced in two of the diodes tends to cancel the distortion in the remaining two diodes. A diode ring will withstand a high overload and has good compression characteristics. However, major problems occur with a diode ring. First, there is considerable signal loss (at least 6 dB). This requires more amplification of the incoming RF signal, as well as higher LO drive power. The IMD is crucially dependent on load termination, and limited isolation is between the input and output. Again, the input impedance is low.

Quad FET mixers have generally good IMD characteristics, will withstand large overloads, and have low noise, but are expensive. The IMD performance depends on the load. Conversion gain (if any), IMD, and noise are not easy to optimize in quad FET mixers.

The SL6440 IC is a balanced mixer where IMD performance is set by current. The LO is properly isolated and a low-power LO can be used. The IC can be used single-ended or with differential drive. The load has little effect on IMD performance. The input impedance is high. It is possible to improve gain by selecting the output impedance. The bandwidth for the SL6440 is limited to 200 MHz. For best noise performance, the noise figure should be 11 dB. The compression point (where further increases in input do not produce corresponding increases in output) is lower than for the best diode rings.

8.3.3 SL6440 characteristics

Figures 8.11, 8.12, 8.13, 8.14, and 8.15 show a basic application circuit, a high-performance receiver mixer, the performance characteristics, frequency response, and pin functions, respectively.

The basic application circuit of Fig. 8.11 is single-ended and does not require any transformer coupling. Because of the voltage drop in the output load resistors, V_{CC1} should be between 2 and 3 V higher than V_{CC2}. The circuit of Fig. 8.11 does not provide gain using the values shown. However, by increasing the load-resistor values (pins 3 and 14) and the value of V_{CC1}, it is possible to get gain with the basic application circuit.

The high-performance circuit of Fig. 8.12 provides gain, but requires interstage coupling transformers. As shown in Fig. 8.13, the gain for the Fig. 8.12 circuit is 10 dB. The circuit characteristics can be improved for transmitter use by increasing the supply voltage to alter the compression point. However, remember that when the SL6440 is adjusted to provide conversion gain, the compression point is reduced.

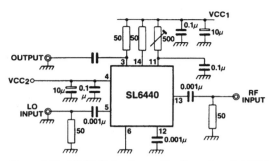

Figure 8.11 SL6440 basic application circuit (*GEC Plessey Semiconductors,* Professional Products IC Handbook, *p. 4-14*).

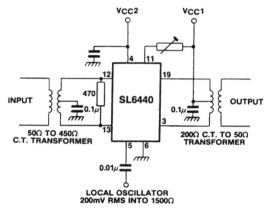

Figure 8.12 SL6440 high-performance receiver mixer (*GEC Plessey Semiconductors,* Professional Products IC Handbook, *p. 4-14*).

Parameter	Performance	Conditions
Sensitivity	15dB S + N/N	1μV EMF, SSB Bandwidth 30MHz, f = 1MHz
IMD 3rd	>-70dB	Input = 142mV EMF each signal 10kHz separation
IMD 2nd	-80dB	Input = 142mV EMF each signal
LO Radiation	-65dBm	Measured in 50Ω at input port
Blocking	100mV EMF	3dB blocking 1μV EMF wanted signal
IF Rejection	30dB	Rejection measured at input port
Input Matching	22dB	Returned loss in a 50Ω system
Gain	10dB	

Figure 8.13 SL6440 performance characteristics (*GEC Plessey Semiconductors,* Professional Products IC Handbook, *p. 4-14*).

Applications for Special-Purpose RF ICs

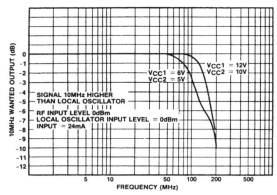

Figure 8.14 SL6440 frequency response (*GEC Plessey Semiconductors*, Professional Products IC Handbook, p. 4-15).

Pin	Function
13 & 12	**Signal input.** These pins may be connected together for best carrier suppression, but should not be DC coupled to any external voltage source or load.
3 & 14	**Open collector outputs.** These are connected to Vcc1, via load resistors, transformer windings, or as required.
4	**Vcc2**, normally about 2 to 3V lower than Vcc1, but for large output swings should be such that the instantaneous voltage at pins 2 and 3 is greater than Vcc2 + 2V.
5	**Local oscillator input.** The LO requirements are for approximately 200mV RMS with an impedance of about 1500Ω.
6	**The negative supply line.**
11	This is the current programming pin. It is connected to Vcc2 via a suitable value of resistor or fed from a current source. The current in at this pin is equal to the current in Pin 2 or Pin 3.

Figure 8.15 SL6440 pin functions (*GEC Plessey Semiconductors*, Professional Products IC Handbook, p. 4-15).

With either the Fig. 8.11 or 8.12 circuit, the IC characteristics can be programmed by the voltage at pin 11, as described in Section 8.3.5.

8.3.4 Thermal considerations for the SL6440

All of the considerations for thermal design in Section 2.17 apply to the SL6440. Here are some additional thoughts on the subject.

The SL6440 chip temperature should not exceed 170°C. As a result of this limitation, the maximum power dissipation depends on the thermal resistance from junction-to-ambient (JA), which is (in free air) about 125°C. This can be reduced with a heatsink (the manufacturer recommends an IERC PEP50AB). Such a heatsink will reduce the JA to about 65°C/W. Heatsinks designed to be attached by thermally conductive epoxy generally have a somewhat higher thermal resistance.

The power dissipation (PD) is given by the following:

$$\text{PD (in mW)} = 2\,I_p\,V_o + V_p\,I_p + 0.75\,(V_{CC2})^2 + 1.5\,V_{CC2}$$

where V_o = dc level at pin 3 or pin 14, V_p = voltage on pin 11, and I_p = programming current in mA. In most applications, V_p is about 2 V.

In many applications, it is possible to run the SL6440 at a low enough dissipation to use a minimum of heatsinking. The circuit of Fig. 8.12 can be operated without a heatsink if temperatures do not exceed +60°C. A simple "glued-on" heatsink, such as an EG & G Wakefield 651B, allows operation to an ambient of +80°C.

8.3.5 Optimizing the SL6440

Use the following information to optimize performance of the circuits in Figs. 8.11 and 8.12.

Gain (G) is given by the expression:

$$G = 20 \log \frac{P_L\,I_p}{56.6\,I_p + 0.0785}$$

for the single-ended (Fig. 8.11) output, and is 6 dB higher where the differential output (Fig. 8.12) is used. R_L is the load resistance (in ohms) from pin 3 or 14 to V_{CC2}. I_p is the programming current (pin 11). The intercept point depends primarily on the value of I_p, and it can be selected using various datasheet graphs. In the absence of a datasheet, adjust the value of I_p (with the 500-Ω resistor at pin 11) until an optimum balance among IMD, noise, dynamic range, and gain is obtained.

When optimizing compression, use output compression, rather than input (because input compression also depends on gain). The voltage on the output pins should be so low that the output transistors in the IC can saturate. This means that the quiescent voltage on these pins must be greater than V_{CC2}. If the output transistors approach saturation, the frequency response might be reduced. This is most noticeable when the SL6440 is used for up conversion. As a minimum, the value of V_{CC1} should be:

$$V_{CC1}\,(\text{min}) = (I_p\,R_L) + V_s + V_{CC2}$$

where I_p is the programming current, R_L is the dc load resistance, and V_s is the peak output voltage.

If V_{CC1} is such that $V_{CC1} = (2\,I_p\,R_L) + V_{CC2}$, then compression will occur at the input of the SL6440.

Applications for Special-Purpose RF ICs   135

8.4 Double-Conversion PLL Detector and RF Mixer

This section describes an IC that provides FM IF, PLL detection, and RF mixing in single chip. The GEC Plessey SL6601 is selected as an example. The IC is a straight-through or single-conversion IF amplifier and detector for FM radio applications. The IC's minimal power consumption makes it ideal for hand-held and remote applications where battery conservation is important.

8.4.1 SL6601 circuit functions and optimization

Figures 8.16, 8.17, and 8.18 show the block diagram, typical application circuit, and electrical characteristics, respectively. The following paragraphs describe the circuit functions and related optimization techniques.

8.4.2 Optimizing the IF amplifier and mixer

The IC can be operated in either Straight-Through mode with a maximum recommended input frequency of 800 kHz, or in Single-Conversion mode (with an input frequency of 50 MHz maximum and an IF of 100 kHz, or 10 times the peak deviation, whichever is larger).

The crystal oscillator frequency can be equal to either the sum or difference of the two IF amplifiers. The exact frequency is not crucial.

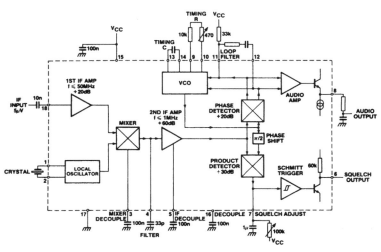

Figure 8.16 SL6601 block diagram (*GEC Plessey Semiconductors, Professional Products IC Handbook, p. 1-155*).

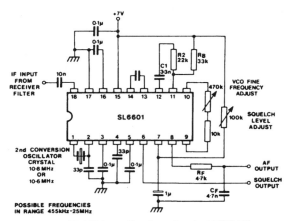

Figure 8.17 SL6601 application circuit (*GEC Plessey Semiconductors*, Professional Products IC Handbook, p. 1-156).

ELECTRICAL CHARACTERISTICS
Test conditions (unless otherwise stated):
Supply voltage Vcc : 7V
Input signal frequency: 10.7MHz, frequency modulated with a 1kHz tone with a ±2.5kHz frequency deviation
Ambient temperature: -30°C to +85°C; IF = 100kHz; AF bandwidth = 15kHz

Characteristic	Value			Units	Conditions
	Min.	Typ.	Max.		
Supply current		2.3	2.7	mA	
Input impedance	100		300	Ω	Source impedance = 200Ω
Input capacity	0.5	2.0	3.5	pF	
Maximum input voltage level	0.5			V rms	At pin 18
Sensitivity	5	2		µV rms	At pin 18 for S + N/N = 20dB
Audio output	35	90	140	mV rms	
Audio THD		1.3	3.0	%	1mV rms input at pin 18
S + N/N	30	50		dB	1mV rms input at pin 18
AM rejection	30	Note 1		dB	100µV rms input at pin 18, 30% AM
Squelch low level		0.2	0.5	V dc	20µV rms input at pin 18
Squelch high level	6.5	6.9		V dc	No input
Squelch hysteresis	1		6	dB	3µV input at pin 18
Noise figure		6		dB	50Ω source
Conversion gain		30		dB	Pin 18 to pin 4
Input gain compression		100		µV rms	Pin 18 to pin 4, 1dB compression
Squelch output load	250			kΩ	
Input voltage range	80	100		dB	At pin 8; above 20dB S + N/N
3rd order intercept point (input)		-38		dBm	Input pin 18, output pin 4
VCO frequency					
Grade 1	85		100	kHz	390pF timing capacitor ⎫
Grade 2	95		110	kHz	390pF timing capacitor ⎬ No input
Grade 3	105		120	kHz	390pF timing capacitor ⎭
Source impedance (pin 4)		25	40	kΩ	
AF output impedance		4	10	kΩ	
Lock-in dynamic range	±8			kHz	20µV to 1mV rms at pin 18
External LO drive level	50		250	mV rms	At pin 2
Crystal ESR			25	Ω	10.8MHz

Figure 8.18 SL6601 electrical characteristics (*GEC Plessey Semiconductors*, Professional Products IC Handbook, p. 1-158).

The circuit is designed to use series-resonant fundamental crystals between 1 and 17 MHz. When a suitable crystal frequency is not available for a particular application, a fundamental crystal at one third of that frequency can be used. However, this results in some degradation performance.

If an external oscillator is used, the recommended oscillator signal level is 70 mV rms. The unused pin should be left open-circuited. The external input is ac coupled through a 0.01-μF capacitor.

A capacitor connected between pin 4 and ground will shunt the mixer output and limit the frequency response of the input signal to the second IF amplifier. A value of 33 pF is recommended when the second IF frequency is 100 kHz (Fig. 8.17). A value of 6.8 pF is recommended for an IF of 455 kHz.

8.4.3 Optimizing the PLL

The PLL detector features a voltage-controlled oscillator with a nominal frequency set by an external capacitor across pins 13 and 14. The value of this capacitor (in pF) is equal to $(40 \pm 7)/f$, where f is the VCO frequency in MHz. Be certain that the free-running frequency of the VCO is as near correct as possible. Both the VCO and limiting-IF amplifier outputs are square waves. It is possible to get a PLL lock with the VCO frequency at some fraction of the IF. For example, if the IF is 100 kHz, the VCO could lock at 150 kHz. This condition produces good S/N characteristics, but poor squelch performance.

The loop filter is connected between pins 11 and 12. A 33-kΩ resistor is also required between pin 11 and V_{CC}. The values for loop-filter resistor R2 and capacitor C1 (Fig. 8.17) must be chosen so that the natural loop frequency and damping factor are suitable for the FM deviation and modulation bandwidth required. Loop filter design is covered further in Section 8.4.8. For most applications, simply use the following values for R_2 and C_1.

Center frequency, kHz	Deviation, kHz	Resistor, kΩ	Capacitor, pF
100	5	6.2	2200
100	10	5.6	1800
455	5	4.7	1500
455	10	3.9	1200

The AF output voltage depends on the percentage of FM deviation. For a given deviation, the output is inversely proportional to the center frequency. Because the noise is constant, the S/N ratio is also inversely proportional to center frequency.

8.4.4 VCO frequency grading

The IC is supplied in three selections of VCO center frequency. In all cases, the frequency is measured with a 390-pF timing capacitor (Fig. 8.18) and with no input signal. The ICs are coded: SL6601C, with a /1, /2, or /3 to indicate VCO center frequency. The center frequency tolerances are:

/1   85 to 100 kHz (or uncoded)
/2   95 to 110 kHz
/3   105 to 120 kHz

8.4.5 Optimizing the squelch

When inputs to the product detector differ in phase, a series of current pulses will flow out of pin 7. This feature can be used for adjustment. Initially, the VCO frequency should be trimmed (using the variable resistance at pin 10) to provide maximum voltage at pin 7. The squelch level is adjusted by means of a variable resistance between pin 7 and V_{CC}. This resistance sets the output S/N ratio at which it is required to mute the output. The capacitor between pin 7 and ground determines the squelch attack time. A value between 10 nF and 10 µF can be chosen to get the required characteristics.

Operation at S/N ratios outside the range of 5 to 18 dB is not recommended. When front-end noise is high (because of very high front-end gain), it is possible that the squelch might not operate. This can be eliminated by reducing front-end gain, and making up for the loss of gain at some other point in the receiver.

The load on the squelch output (pin 6) should not be less than 250 kΩ. Reduction of the load below 250 kΩ might result in hysteresis problems in the squelch circuit. Figure 8.19 shows how an external PNP transistor can be used to increase hystersis. The use of capacitors greater than 1000 pF from pin 6 to ground is not recommended.

8.4.6 Outputs

High-speed data outputs can be taken directly from pins 11 and 12. However, for typical audio applications, use the output at pin 8. The RC network consisting of 4.7 kΩ and 4.7 nF is recommended to restrict the audio bandwidth.

8.4.7 Layout and VCO adjustment

The SL6601 has no crucial layout problems. A possible exception is where the IC is used in Straight-Through mode, where input compo-

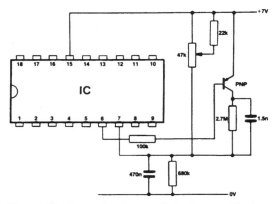

Figure 8.19 Increasing hysteresis in the squelch circuit (*GEC Plessey Semiconductors,* Professional Products IC Handbook, *p. 1-156*).

nents and circuits should be isolated from the VCO components. This will prevent the VCO from attempting to lock to itself, resulting in poor S/N ratios.

The recommended method of VCO adjustment is with a frequency counter at pin 9, and no signal at the IF input (pin 18). The counter impedance should be high so as not to overload the VCO circuit.

8.4.8 Optimizing the loop filter

Although the values for the loop filter (R2 and C1 in Fig. 8.17) are not crucial, and the values given in Section 8.4.3 are suitable for most typical applications, there might be occasions where it is necessary to select values for the filter. The major concern in loop-filter design is that the loop bandwidth is not so low as to allow unlocking of the loop when modulation occurs. Another primary concern is that the damping factor can be chosen for maximum flatness of frequency response or for minimum noise bandwidth. In the case of the SL6601, damping-factor values between 0.5 and 0.8 are realistic, with 0.5 providing the minimum noise bandwidth. The remainder of this section is devoted to loop-filter design for the SL6601, but the information can be applied (in general) to most PLL loop filters.

The design starts with an arbitrary choice of f_n, the natural loop frequency. By setting this at slightly higher than the maximum modulation frequency, f_m, the noise rejection can be slightly improved. The ratio f_m/f_n (highest modulating frequency to loop frequency) can then be evaluated.

The graph of Fig. 8.20 shows that the damping factor is established by:

$$\theta e\, f_n$$

where θ = peak phase error, f_n = loop natural frequency, and d_f = maximum deviation of the input signal.

Because f_n and d_f are known, θe can be easily calculated. Values for θe should be chosen so that error in phase is between 0.5 and 1 radian. This is because the phase detector limits at ±6.28/2 radians and is nonlinear approaching these points.

Using a very small peak phase error means that the output from the phase detector is low; thus, it impairs the S/N ratio. As a result, the choice of a compromise value, and 0.5 to 1 radian is used. If the value of θe achieved is far removed from this value, a new value of f_n should be chosen and the process repeated.

With f_n and D established, the time constants are derived from:

$$t_1 + t_2 = \frac{K_o\, K_D}{(6.28\, f_n)^2}$$

and

$$t_2 = \frac{D}{3.14\, f_n} - \frac{1}{K_o\, K_D}$$

$K_o\, K_D$ is $0.3 f_o$, where f_o is the operating frequency of the VCO. t_1 is fixed by the capacitor and an internal 20-kΩ resistor. t_2 is fixed by the capacitor and external. As a result,

$$C = \frac{t_1}{20 \times 10^3}$$

and

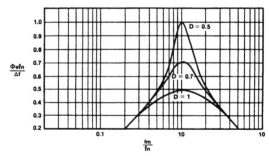

Figure 8.20 Damping factor for PLL loop filter (*GEC Plessey Semiconductors,* Professional Products IC Handbook, *p. 1-156*).

$$R_{ext} = \frac{t_2 \times 20 \times 10^3}{t_1}$$

So that standard values can be used, it is better to establish a value for C and then use the next lowest standard value. For example, if the calculated value for C is 238 pF, use 220 pF. Also, it is better to widen the loop bandwidth, rather than narrow the bandwidth.

The value of R_{ext} should be rounded up by a similar process. However, it is better to use the next larger standard value for the external resistor. This is because loop bandwidth is proportional to $(R_{ext}) -\frac{1}{2}$, but the damping factor is proportional to R. That is, damping factor is increasing more quickly, which provides a more level response.

8.4.9 Example of loop filter design

In the following example of PLL loop-filter design, it is assumed that the FM signal has a deviation of 10 kHz and a maximum modulation of 5 kHz. The VCO frequency is 200 kHz, the loop natural frequency is 6 kHz, and $D = 0.5$.

With an f_m/f_n of 5/6 and a D of 0.5, the graph of Fig. 8.20 shows:

$$\frac{\theta e f_n}{d_f} = 0.85$$

$$\theta e = \frac{0.85\, d_f}{f_n} = \frac{0.85 \times 10}{6} = 1.4 \text{ rad}$$

This is too large (greater than 1), so increase f_n to 10 kHz.

$$\frac{f_m}{f_n} = \frac{0.5\, \theta e f_n}{d_f} = 0.45$$

$$\theta e = \frac{0.45 \times 10}{10} = 0.45$$

This is too low (less than 0.5), so set $f_n = 7.5$ kHz.

$$\frac{f_m}{f_n} = \frac{5}{7.5} = 0.666$$

$$\frac{\theta e f_n}{d_f} = 0.666$$

$$\theta e = \frac{0.666 \times 10}{7.5} = 0.888 \text{ rad}$$

$$t_1 + t_2 = \frac{K_o K_D}{(6.28 f_n)^2}$$

$$K_o K_d = 0.3 f_o$$

where f_o is the VCO frequency (200 kHz).

$$t_1 + t_2 = \frac{0.3 \times 200 \times 10^3}{(6.28 \times 7.5 \times 10^3)^2} = 27 \text{ }\mu\text{s}$$

$$t_2 = \frac{D}{6.28 f_n} - \frac{1}{K_o K_D}$$

$$= \frac{0.5}{3.14 \times 7.5 \times 10^3} - \frac{1}{0.3 \times 200 \times 10^3}$$

$$= 4.6 \text{ }\mu\text{s}$$

$$t_1 = 22.4 \text{ }\mu\text{s}$$

$$C = \frac{t_1}{20 \times 10^3} = \frac{22.4 \times 10^{-6}}{20 \times 10^3} = 1.12 \text{ nF (use 1 nF)}$$

$$R = \frac{t_2}{t_1} \times (20 \times 10^3)$$

$$= \frac{4.6}{22.4}$$

$$= 4.1 \text{ k}\Omega$$

Chapter 9

Optimizing Frequency Synthesizers

This chapter is devoted to optimizing frequency synthesis (FS) circuits, such as covered in Section 1.6.2. The GEC Plessey SP8853 is selected as an example. This IC uses a form of *multimodulus division* (also known as *pulse swallowing*) to optimize FS performance. Before getting into the IC characteristics and how the IC can be optimized in practical FS circuits, first review multimodulus division and some common synthesizer problems.

9.1 Loop Bandwidth

Loop bandwidth is an important characteristic in PLL design. This is because loop bandwidth determines such parameters as lock-up time, noise, and modulation capability. In single-loop synthesizers, loop bandwidth is generally made as wide as possible. However, there are conflicting requirements and single-loop synthesizers are not always practical.

If the loop is slewed at too high a rate, then a longer lock-up time might result because of overshoot. In extreme cases, the loop will become unstable because the VCO frequency will sweep too quickly.

Phase noise of the VCO inside the loop bandwidth will be reduced by the loop. Outside the bandwidth, phase noise is unaltered. This is because of the PLL characteristics. When signals are injected into the loop, a PLL acts as a low-pass filter for signals inside of the bandwidth, and as a high-pass filter for signals outside of the loop bandwidth.

To understand this condition, consider modulation of the VCO at very low frequencies. The output of the phase detector will be a low-frequency signal of phase such as to remove the VCO modulation.

When the modulation frequency increases, the error component of the phase detector output is not passed by the loop filter, so the modulation is not removed by the loop. Notice that the modulation is phase modulation (PM) up to the filter breakpoint and frequency modulation (FM) thereafter.

The phase noise of the reference oscillator will add to the VCO noise at frequencies inside the loop bandwidth; this effect also influences the choice of loop bandwidth. For example, a loop with a 5-kHz loop bandwidth operating at the typical cellular frequency of 900 MHz, with a reasonable 5-MHz crystal-oscillator noise floor (for example, -125 dBc/Hz at a 1-kHz oscillator) would have a noise power of about -80 dBc/Hz at the 1-kHz offset of the final frequency.

When a high phase-detector gain is used with a noisy oscillator, or with a high value of K_v (K_v = VCO sensitivity in Hz/volt), it is possible that the phase detector will be driven outside the phase window. This will lead to instability.

Modulation of the PLL can occur inside or outside of the loop bandwidth. Modulation outside of the loop bandwidth requires that the bandwidth be less than the lowest modulating frequency; the amount of modulation will vary over the frequency range when K_v varies. A number of techniques are used to minimize the variation in modulation sensitivity. The easiest to use is a separate modulation diode. Typically, deviation will remain reasonably constant over a wide range (using a diode).

Modulation outside of the loop bandwidth causes a signal to appear at the phase detector output. This signal corresponds to the phase error between the reference frequency and the divided VCO. If the phase error drives the phase detector outside of the phase window, reference-frequency sidebands can occur, and the loop can become unlocked.

Modulation inside of the loop bandwidth avoids some of these problems, but care must be taken to ensure that the reference-frequency sidebands do not become large. Also, the wideband noise of the phase detector and loop filter can cause problems when K_v is high with modulation inside of the loop.

Modulation of the reference oscillator is a possible technique when modulating inside of the loop bandwidth. However, all modulation inside of the loop bandwidth produces phase modulation, rather than frequency modulation. Also, there are limits on the frequency deviation and modulation frequency that can be accepted without the loop becoming unlocked. Generally, the modulation frequency must be much less than the loop bandwidth. For this reason, the trend is to use modulation outside of the loop bandwidth—especially when the PLL is used at wireless frequencies (800 to 900 MHz) and above.

9.2 Multimodulus Division

Figure 9.1 shows the *direct-division* form of PLL frequency synthesizer. Such FS circuits have problems when a fully programmable divider is required to operate at higher frequencies. Figure 9.2 shows a PLL FS circuit with *fixed prescaling,* where the phase-comparison frequency of the Fig. 9.1 circuit is reduced by the factor f/N. This lower frequency requires a lower bandwidth in the PLL, thus making the circuit more susceptible to microphonics and requiring a longer lock-up time.

The alternatives to fixed direct division, or fixed prescaling, are *mixing* (Fig. 9.3) or *multimodulus division* (pulse swallowing), as shown in Fig. 9.4. The use of mixers requires great care in the choice of frequencies to minimize the effects of undesired frequency products. Mixers are, however, widely used—even though they are generally more complicated (susceptible to layout problems, parts selection after testing, etc.).

The multimodulus divider system of Fig. 9.4 requires three functions, in addition to the VCO, loop filter, phase detector, and reference-frequency divider used in all PLL systems. These three functions include:

- A two-modulus divider, which will be divided by one of two numbers N or $N + 1$ (10/11, 64/65 etc.).

- An A counter, which is programmable and the output of which controls the modulus of the divider.

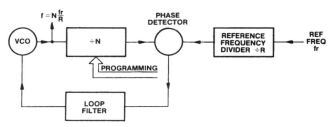

Figure 9.1 Direct-division PLL (*GEC Plessey Semiconductors, Professional Products IC Handbook, p. 4-38*).

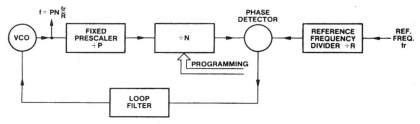

Figure 9.2 Fixed-prescaling PLL (*GEC Plessey Semiconductors,* Professional Products IC Handbook, *p. 4-38*).

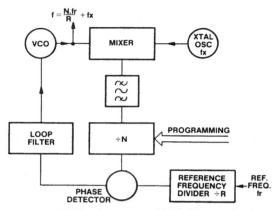

Figure 9.3 PLL with mixer (*GEC Plessey Semiconductors,* Professional Products IC Handbook, *p. 4-38*).

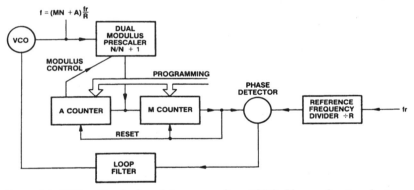

Figure 9.4 PLL with dual-modulus prescaling (*GEC Plessey Semiconductors,* Professional Products IC Handbook, *p. 4-39*).

- An M counter, which is programmable and clocked in parallel with the A counter, and the output of which resets both itself and the A counter.

The counters can count down to zero from the programmed input, or count up from zero.

The A counter is programmed to a smaller number than the M counter. Assuming that the counters are empty, the system starts with the divider ($N/N + 1$) dividing by $N + 1$. This continues until the A counter reaches its programmed value. At that point, the divider divides by N until the M counter is full. Because the M counter has received A pulses, the M counter overflows after $(M - A)$ pulses, corresponding to $N(M - A)$ input pulses to the divider.

The total division ratio (P) is given by:

$$P = (N + 1)A + N(M - A) = NM + A$$

Obviously, A must be equal to or less than M for the system to work. For every possible channel to be available, the minimum total divide ratio is $N(N - 1)$ and the maximum total divide ratio is $M(N + 1)$. A_{max} should be equal to or greater than N.

Although this appears to be simple, problems can occur when designing a multimodulus PLL system. *Loop delay* is probably the major problem. Consider the counter chain at the instant that the $(N + 1)$th pulse appears at the two-modulus divider input. After some time (t_{p1}), the output produces a pulse which clocks the A and M counters. Assume that the A counter is filled by the pulse and, after a time t_{p2} (determined by the propagation delay of the A counter), an output is produced to set the dual-modulus divider ratio to N.

After a set-up time (t_s), the dual-modulus divider will divide by N. However, if $t_{p1} + t_{p2} + t_s$ is greater than N cycles of input frequency, the divider will not be set to divide by N until after N pulses have appeared, and the system will fail. As a result, the total loop delay must be less than N divided by the input frequency. Design in this area is crucial. Worst-case tolerances must be used for proper reproducibility and reliability of design, especially under temperature and voltage extremes. The value of N must also be large enough that the output frequency from the divider does not exceed the maximum input frequency of the following circuits.

Although the minimum value for N is set by these theoretical limits, the actual value of N (above the minimum) is usually determined by ease of programming. For example, assume that the synthesizer has a 25-kHz phase-comparison frequency and 25-kHz channel spacing, using a 40/41 divider.

At 156 MHz:

$$P = \frac{156}{0.025} = 6240$$

Therefore,

$$NM + A = 6240$$

$$40M + 0 = 6240 \ (A = \text{zero for the lowest channel})$$

$$M = 156$$

In general, where:

$$fN = 1 \text{ or } 10 \text{ or } 100$$

$$M = f, \frac{f}{10}, \frac{f}{100}, \text{etc.}$$

and similarly for binary-divide ratios.
The choice of a prescaler is, therefore, fixed by:

1. Total allowable loop delay, where N divided by the input frequency is greater than the total delays.
2. Output frequency with the input frequency band.
3. Programming ease.

The *reference-frequency division ratio* (R) is another important factor in PLL synthesizer design. The value of R is set by the input frequency and the phase-comparison frequency, where R = *input frequency ÷ phase-comparison frequency*. Higher input frequencies require greater power and offer lower stability. Lower frequencies (below 4 MHz) generally require larger crystal-case sizes.

The value of R must be an even number. For example, assume that the input frequency is 10.245 MHz (from a local oscillator). An R value of 4098 would produce a comparison frequency of 2.5 kHz.

9.3 SP8853 Characteristics

Figures 9.5 and 9.6 show the block diagram and electrical characteristics, respectively. The SP8853 is a low-power single-chip synthesizer intended for wireless/RF applications in the 1.3- to 1.5-GHz range, and contains all the elements (apart from the loop amplifier) to fabricate a PLL FS loop. The IC is serially programmable by a three-wire data highway, and contains three independent buffers to store one reference divider word and two local-oscillator divider words.

A digital comparator, with two charge pumps (programmable in phase and gain), is provided to improve lock-up performance. The preset tandem operation of the charge pumps can be overwritten or the comparison frequencies can be switched to output ports under the control of the divider word. The dual-modulus ratio (and so the operating range) is also programmable through the same word. A Power-Down mode is incorporated as a battery economy feature. The IC is specified to 1.5 GHz at +85°C and to 1.3 GHz at +125°C in a DG package. In the HC package, the IC is specified to 1.3 GHz at +85°C and +125°C.

9.4 Prescaler and *A* and *M* Counters

The programmable divider chain contains a dual-modulus front-end prescaler, an A counter (which controls the dual-modulus ratio), and an M counter (which performs the bulk multimodulus division). The

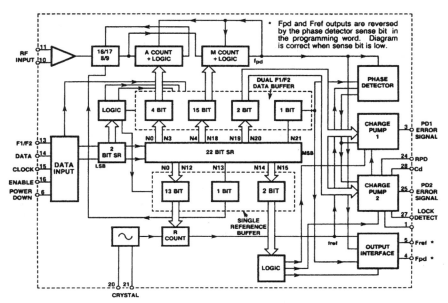

Figure 9.5 SP8853 block diagram (*GEC Plessey Semiconductors,* Professional Products IC Handbook, *p. 2-4*).

divider has a division ratio of $MN + A$, and a minimum integer steppable division ratio of $N(N - 1)$.

The dual-modulus front-end prescaler is a dual N ratio device that is capable of being switched between 16/17 and 8/9 ratios. The controlling A counter is of 4-bit design, permitting a maximum count sequence of 15 ($2^4 - 1$). The A counter sequence begins with the start of the M counter sequence, and stops when a preloaded number of cycles is reached. While the A counter is counting, the dual-modulus prescaler is held in the $N + 1$ mode, and then shifts back to the N mode when the sequence is complete.

The M counter is a 15-bit asynchronous divider, which counts with a ratio set by a control word. In both A and M counters, the controlling data from the F_1/F_2 buffer is loaded in sequence with every M count cycle. The N ratio of the dual-modulus prescaler is selected by a 1-bit word in the reference-divider buffer. When a ratio of 8/9 is selected, the A counter requires only 3 programming bits.

9.5 Reference Source and Divider

The reference source is obtained from an on-board oscillator, which is frequency controlled by an external crystal at pins 20/21. The oscillator can also function as a buffer amplifier, allowing the use of an external refer-

ELECTRICAL CHARACTERISTICS
Test conditions (unless otherwise stated)
Supply Voltage: V_{CC} = +4.75V to +5.25V
Temperature: A Grade T_{amb} = -55°C to +125°C
            B Grade T_{amb} = -40°C to +85°C

Characteristic	Pin	Value			Units	Conditions
		Min.	Typ	Max.		
Supply Current	8,9		33	40	mA	
	18,23		4.5	6	mA	
Supply Current in Power Down Mode	8					
Input Sensitivity	10,11					See Figs. 4a and b
Input Overload	10,11					See Figs. 4a and b
RF Input Division Ratio	10,11,4	256		524287		With 16/17 selected
		56		262143		With 8/9 selected
Comparison Frequency	4,5			5	MHz	
Reference Oscillator Input Frequency	20,21	4		20	MHz	
External Reference Input Voltage	20	10		500	mVrms	
Reference Division Ratio	20,5	1		8191		
Data Clock Repetition Rate t_{rep}	15			1	μs	See Fig. 5
Minimum Set up Time t_s	14,15	50			ns	See Fig. 5
Data Input High	14	0.6Vcc		Vcc	V	
Low	14	Vee		0.3Vcc	V	
Clock Input High	15	0.6Vcc		Vcc	V	
Low	15	Vee		0.3Vcc	V	
Data Enable High	16	0.6Vcc		Vcc	V	
Low	16	Vee		0.3Vcc	V	
F1/F2 Input High	13	0.6Vcc		Vcc	V	F1 buffer selected
Low	13	Vee		0.3Vcc	V	F2 buffer selected
Power Down Input High	6	0.6Vcc		0.9Vcc	V	
Low	6	Vee		0.3Vcc	V	
F1/F2 Input Current	13			5	μA	V Pin 13 = 5.0V
Power Down Input Current	6			5	μA	v Pin 6 = 4.5V
RPD External Resistance	24	68		330	kΩ	
Lock Detect Output Voltage 'in lock'	27			1	V	I pin 27 = 1mA
Lock Detect Switching Voltage High	25	2.7			V	Vcc = 5V
Low	25			2.3	V	Vcc = 5V
Fpd and Fref Output Voltage Swing			0.9		V	Vcc = 5V. External pull down may be required

Figure 9.6 SP8853 electrical characteristics (*GEC Plessey Semiconductors,* Professional Products IC Handbook, *p. 4-6*).

ence source. In this mode, the source is ac-coupled into the oscillator transistor base on pin 20.

The oscillator output is coupled to a programmable reference divider. The output from the divider is the reference for the phase detector. The reference divider is a fully programmable 13-bit asynchronous device, and it can be set to any division ratio between 1 and 8191. The actual division ratio is controlled by a data word stored in the internal reference buffer.

9.6 Phase Comparator

The IC has a digital phase comparator that feeds two charge-pump circuits. Charge-pump 1 has preset currents, and charge-pump 2 has a current level set by an external resistor. The current is multiplied by a factor determined by the F_1 or F_2 word, as follows:

F_1 or F_2 word $G_2 \quad G_1$		Charge-pump 1 current	Charge-pump 2 multiplier
0	0	50 µA	1
1	0	75 µA	1.5
0	1	125 µA	2.5
1	1	200 µA	4

Note that the charge-pump 2 current is set by the R_{PD} resistor at pin 24, as shown in Fig. 9.7. The pin 24 current is set by $(V_{CC} - 1.5)/R_{PD}$.

A lock-detect circuit is connected to the charge-pump 2 output. When the voltage level at pin 25 is between about 2.25 and 2.75 V, pin 27 goes low and charge-pump 1 is disabled, depending on the PD1 and PD2 programming bits:

PD2	PD1	
0	0	F_{ref} and F_{pd} outputs off, charge-pump 1 and 2 on.
0	1	F_{ref} and F_{pd} outputs on, charge-pump 1 off, charge-pump 2 on.
1	0	F_{ref} and F_{pd} outputs off, charge-pump 1 disabled by lock-detect, charge-pump 2 on.
1	1	F_{ref} and F_{pd} outputs on, charge-pump 1 disabled by lock-detect, charge-pump 2 on.

The output signals from the reference counter and M counter are available on pins 4 and 5 when programmed by the reference-programming word (PD1 and PD2). An external phase detector can be connected to pins 4 and 5, and can be used independently or in conjunction with the on-chip phase detector.

To allow for control-direction changes introduced by design of the control loop, a programming bit in the F_1/F_2 programming word interchanges the inputs to the on-chip phase detector and reverses the functions on pins 4 and 5.

9.7 Data Entry and Storage

Figure 9.8 shows the data formats for the SP8853. Table 1 (referred to in Fig. 9.8) is the F_1/F_2 word table in Section 9.6. Table 4 (referred to in Fig. 9.8) is the PD1/PD2 table in Section 9.6.

As shown in Fig. 9.8, the data section of the SP8853 consists of a data-input interface, an internal data-shift register, and three internal data buffers. Data bits are entered to the data-input interface by a three-wire data highway with data, clock, and chip-enable inputs. The

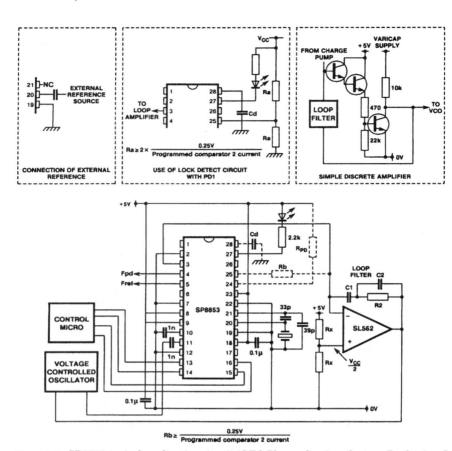

Figure 9.7 SP8853 typical application circuit (*GEC Plessey Semiconductors,* Professional Products IC Handbook, *p. 2-8*).

input interface then routes this data to a 24-bit shift register with bus connections to three data buffers. Data bits entered via the serial bus are transferred to the appropriate data buffer on the negative transition of the chip-enable input, according to the two final data bits (the MSB of the data is entered first):

2-bit SR contents	Buffer loaded
00	F_1
10	F_2
01	Active A*
11	Reference

*Transfer of *A* counter bits into a buffer controlling the programming counter.

Optimizing Frequency Synthesizers 153

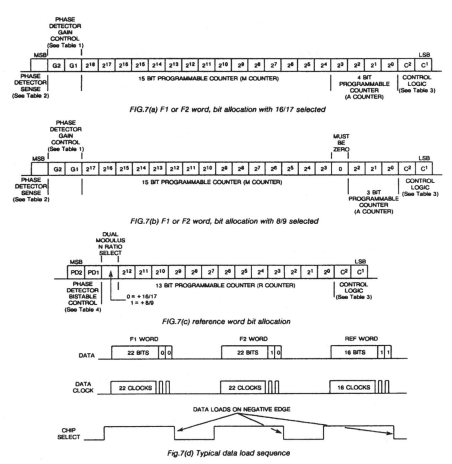

Figure 9.8 SP8853 data formats (*GEC Plessey Semiconductors,* Professional Products IC Handbook, *p. 2-9*).

The dual F_1/F_2 buffer can receive two 22-bit words and it controls the programming divider A and M counters using 19 bits, the phase-detector gain with two bits, and the phase-detector sense with one bit:

Sense bit	Pins 3 and 25
0	Current source
1	Current sink

Note that a fourth input from the synthesizer control system selects the active buffer.

The third buffer contains only 16 bits. Thirteen bits are used to set the reference-counter division ratio and 2 bits are used to control the phase-comparator enable logic. The remaining bit sets the dual-modulus prescaler N ratio.

The data words can be entered in any individual multiple sequence and the shift register can be updated while the data buffers retain control of the synthesizer with the previously loaded data. This enables four unique data words to be stored in the IC, with three in the data buffers and a fourth in the shift register (during the time that the chip is enabled). The F_1 word can also be updated while the F_2 word is controlling the programmable divider, and vice versa.

The dual F_1/F_2 buffer enables the IC to be toggled between two frequencies using the F_1/F_2 select input at a rate determined by the comparison frequency. This also enables random frequency hopping at a rate that is determined by a byte-load period because the loop can be locked to F_1, while F_2 is updated by entering new data via the shift register. (The F_1/F_2 input must be high to select F_1.)

An F_1 or F_2 update cycle consists of a byte that contains 24 bits. The reference byte contains 18 bits. The IC requires 3 bytes, each with a chip-select sequence, totalling 66 bits to fully program the IC.

When the dual-modulus counter (A counter) is set to $+8/9$, the data bits required to set the counter are reduced by one bit. An unused bit is left in the 22-bit F/F_2 buffer. This unused bit must always be set to zero when the $+8/9$ mode is required.

The data-entry and storage registers are always powered up, making it possible to enter data when the IC is in the power-down state.

9.8 Optimizing SP8853 Circuits

The basic application circuit shown in Fig. 9.7 is designed for operation at frequencies up to 1.5 GHz (beyond the usual wireless range), so good RF-design techniques must be used. As a minimum, this includes a ground plane and suitable high-frequency capacitors at the RF input and for power-supply decoupling.

The RF input should be coupled to either pin 10 or 11, with the other decoupled to ground. The reference oscillator is of the conventional Colpitts type (Section 2.16) with two capacitors required to provide a low-impedance tap for the feedback signal to the transistor emitter. Typical values are shown in Fig. 9.7. The amplitude should be kept below 0.5 V rms to avoid forward biasing the transistor collector-base junction.

In some systems, it is useful to have an indication of phase lock. The output from pin 27 goes low when the output of charge-pump 2 is between 2.25 and 2.75 V, and can be used to operate an LED to give visual indication of phase lock. As an alternative, a pull-up resistor can be connected to V_{CC} and the output can be used to signal the control microprocessor that the loop is locked, thus speeding up system operation. The output current available from pin 27 is limited to 1.5 mA. If this current is exceeded, the logic-low level will be uncertain.

Although the circuit of Fig. 9.7 is simple (with a minimum component count), the circuit can be used directly in many applications. Charge-pump 1 on pin 3 is used to drive the loop amplifier, which provides the control voltage for the local oscillator. When charge-pump 1 is used in this mode, the (PD1 and PD2 bits in the reference-programming word (Fig. 9.8) must be set to enable charge-pump 1 continuously (PD1 and PD2 both at 0). This application could also use the charge-pump 2 output on pin 25. If a higher phase-comparator gain is required, pins 3 and 25 could be connected in parallel to use the combined output current from both charge-pumps.

The lock-detect circuit can be programmed to disable charge-pump 1 automatically (PD1 = 0 and PD2 = 1, or both PD1 and PD2 = 1). This feature can be used to reduce the system lock-up time by connecting the charge-pump outputs in parallel to the loop amplifier with resistor R_b in series with charge-pump 2. This connection allows a relatively high current to be used from charge-pump 1 to provide short lock-up time and a low current to be set on charge-pump 2 (resulting in low reference-frequency sidebands). The degree of lock-up time improvement depends on the ratio of charge-pump 1 to charge-pump 2 currents.

When the loop is out of lock, both charge pumps are enabled and feed current to the loop amplifier to bring the oscillator to phase lock. The current from charge-pump 2 produces a voltage drop across R_b, allowing operation of the lock-detect circuit and enabling charge-pump 1. The resistor must be chosen to provide a voltage drop greater than 0.25 V at the current level that is programmed for charge-pump 2. When phase lock is achieved, there is no charge-pump current and the voltage at pin 25 is equal to that on the virtual ground point of the loop amplifier (about 2.5 V). This disables charge-pump 1.

Charge-pump 1 should not be left open-circuited when enabled because this prevents correct operation of the phase detector. The output on pin 3 should be biased to half supply ($V_{CC}/2$) with a pair of 4.7-kΩ resistors connected between supplies.

When charge-pump 2 is used to drive the loop amplifier, the lock-detect circuit will provide an out-of-lock indication only when large frequency changes are made, or when a frequency outside of the range of the local oscillator is programmed. At other times, the loop-amplifier input is maintained at 2.5 V by the action of the loop-filter components. Again, resistor R_b connected between pin 25 and the loop amplifier (producing a voltage drop greater than 0.25 V at the programmed charge-pump current) allows sensitive out-of-lock detection.

When phase-lock detection is required using comparator 1 only (inset of Fig. 9.7), charge-pump 2 output (pin 25) should be biased to 2.5 V using two equal-value resistors (R_a) across the supply. The values should be chosen to provide a voltage change greater than 0.25 V at the programmed comparator 2 charge-pump current. A small capacitor (C_d) connected from pin 28 to ground can be used to reduce the chatter of the lock-detect circuit.

An amplifier is required to convert the current pulses from the phase comparator into a voltage of suitable magnitude to drive the chosen VCO. The choice of amplifier must be determined by the voltage swing required at the VCO to achieve the necessary frequency range. In most cases, an op amp will be used to provide the essential characteristics of high input impedance, high gain, and low output impedance required for this application.

Although it is expected that an op amp will be used in most cases, a simple discrete-component amplifier can be used. Such a design is shown in the inset of Fig. 9.7. This arrangement can be particularly useful when the minimum VCO control voltage must be close to ground and where negative supplies are inconvenient. (This discrete form of amplifier is not suitable with charge-pump 2 when the lock-detect circuit is required.)

When an op amp is used in the inverting mode (as shown in Fig. 9.7), the charge-pump output is connected directly to the virtual ground point, and will therefore operate at a voltage similar to that set on the noninverting input of the op amp. Normally, this operating point should be set at half supply ($V_{CC}/2$) using two equal-value resistors (R_x). If necessary, this voltage can be set 1 V higher or lower than half supply without detrimental effect. When the lock-detect function is required on charge-pump 2, the noninverting input will be at half supply.

The digital phase comparator and charge pump used on the SP8853 produces bidirectional current pulses to correct errors between the reference and VCO divider outputs. When synchronization is achieved, no further output from the charge pump should be required (in theory). In practical applications, the capacitors forming the loop filter around the amplifier will gradually discharge because of leakage currents—especially the input current of the amplifier. This action requires further outputs from the charge pump to restore the charge. The effect of continuous correction for the local-oscillator frequency is to frequency modulate the VCO. This results in sidebands at the reference frequency. To minimize this effect, use an amplifier with low input-bias current.

9.9 Optimizing the Loop Filter

The loop filter shown in Fig. 9.7 can either be second order or third order. Figure 9.9 shows a standard form second-order loop filter. In practical applications, an additional time constant (shown dotted in

Fig. 9.9) is often added to reduce noise from the amplifier. In addition, any feed-through capacitor or local decoupling at the VCO will be added to C_2. These additional components form a third-order loop. If the values are chosen correctly, the additional filtering provided can considerably reduce the level of reference-frequency sidebands and noise without affecting loop settling time. The remainder of this chapter is devoted to calculating the values of second-order and third-order loop filters for the SP8863. The same basic calculations can be applied to loop filters for other frequency-synthesizer ICs.

9.10 Second-Order Loop Filter Calculations

Two basic equations are required for the second-order filter shown in Fig. 9.9. The equations are required to determine the time constants t_1 and t_2, where:

$$t_1 = C_1 R_1$$
$$t_2 = C_1 R_2$$

The equations are:

$$t_1 = \frac{K_1 K_2}{(6.28n)^2 N}$$

$$t_2 = \frac{2DF}{6.28n}$$

where K_1 = Phase-detector gain factor in V/radian
K_2 = VCO gain factor in radians second/volt
N = Division ratio from the VCO to reference frequency
n = Natural loop bandwidth
DF = Damping factor (normally 0.7071)

The SP8853 phase comparator is a current source, rather than a conventional voltage source, and has a gain factor specified in μA/radian.

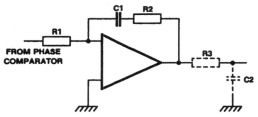

Figure 9.9 Standard-form second-order loop filter (*GEC Plessey Semiconductors,* Professional Products IC Handbook, *p. 2-12*).

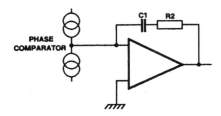

Figure 9.10 Modified second-order loop filter (*GEC Plessey Semiconductors*, Professional Products IC Handbook, *p. 2-12*).

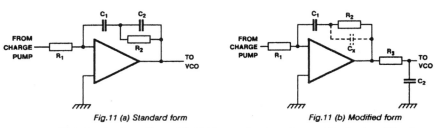

Figure 9.11 Third-order loop filters (*GEC Plessey Semiconductors*, Professional Products IC Handbook, *p. 2-13*).

Because the equations deal with a filter where R1 is feeding the virtual ground point of an op amp from a voltage source, R1 sets the input current to the filter. This is similar to the circuit shown in Fig. 9.10, where a current-source phase comparator is connected directly to the virtual ground point of the op amp.

The equivalent voltage gain of the phase comparator can be calculated by assuming a value for R1 and calculating a gain in volts/radian that would produce the set value. The phase comparator used in the SP8853 is linear over a range of 6.28 radians. Thus, the phase-comparator gain is given by:

$$\frac{\text{phase-comparator current setting (}\mu\text{A/radian)}}{6.28}$$

Assuming a value of 1 kΩ for R1, and 50 μA (Section 9.6) for the phase-comparator current setting, the phase-comparator gain is:

$$\frac{50 \ \mu\text{A} \times 1000}{6.28}$$

These values can be inserted into the t_1 and t_2 equations to find the values of C_1 and R_2, as described in the following example. Calculate the values for a second-order loop filter with the following parameters:

Frequency to be synthesized	800 MHz
Reference frequency	100 kHz
Division ratio, N	$\dfrac{800 \text{ MHz}}{100 \text{ kHz}} = 8000$
Natural loop frequency, n	500 Hz
VCO gain factor, K_2	6.28×10 MHz/Volt
Damping factor, DF	0.7071
Phase-comparator current setting	50 μA (Section 9.6, $G_1/G_2 = 0$)

Assuming that $R_1 = 1$ kΩ, then the equivalent phase-comparator gain (K_1) in V/radian is:

$$\frac{50 \text{ μA} \times 1000}{6.28} = 0.00796 \text{ V/radian}$$

With K_1 at 0.00796 and K_2 given at 6.28×10 MHz/V, then t_1 is:

$$\frac{0.00796 \times 6.28 \times 10 \text{ MHz}}{(6.28 \times 500)^2 \, 500} = 6.334 \times 10^{-6}$$

With DF at 0.7071 and the natural loop frequency at 500 Hz, then t_2 is:

$$\frac{2 \times 0.7071}{6.28 \times 500} = 4.50 \times 10^{-4}$$

Because $t_1 = C_1 R_1$, C_1 is:

$$\frac{6.334 \times 10^{-6}}{1000} = 6.33 \text{ nF}$$

Because $t_2 = C_1 R_2$, R_2 is:

$$\frac{4.50 \times 10^{-4}}{6.33 \times 10^{-9}} = 71.090 \text{k}\Omega$$

Use the standard value of 71 kΩ.

9.11 Third-Order Loop Filter Calculations

The third-order loop filter is normally as shown in Fig. 9.11A. Figure 9.11B shows the circuit redrawn to use an RC time constant after the amplifier. This allows any feed-through capacitance on the VCO line to be included in the loop calculations. Where the modified form in Fig. 9.11B is used, it is helpful to connect a small capacitor, C_x (typically 100 pF, shown dotted), across R2 to reduce sidebands caused by the amplifier being forced into nonlinear operation by the phase-comparator pulses.

Three basic equations are required for the third-order filter shown in Fig. 9.11. The equations are required to determine the time constants t_1, t_2, and t_3, where:

For Fig. 9.11A

$$t_1 = C_1 R_1$$
$$t_2 = R_2 (C_1 + C_2)$$
$$t_3 = C_2 R_2$$

For Fig. 9.11B

$$t_1 = C_1 R_1$$
$$t_2 = C_1 R_2$$
$$t_3 = C_2 R_3$$

The equations are:

$$t_1 = \frac{K_1 K_2}{N (6.28n)^2} \left| \frac{1 + (6.28n)^2 (t_2)^2}{1 + (6.28n)^2 (t_2)^2} \right| \frac{1}{2}$$

$$t_2 = \frac{1}{(6.28n)^2 (t_3)}$$

$$t_3 = \frac{-\tan \theta o + \dfrac{1}{\cos \theta o}}{6.28n}$$

where K_1 = Phase-detector gain factor in V/radian
K_2 = VCO gain factor in radians second/volt
N = Total division ratio from the VCO to reference frequency
n = Natural loop bandwidth
θo = The phase margin (normally 45°)

As in the case of the second-order filter example (Section 9.10), a value for R1 can be assumed and an equivalent gain (K_1), in V/radian can be calculated from:

$$\frac{\text{phase-comparator current setting (μA/radian)}}{6.28}$$

Assuming a value of 1 kΩ for R_1 and 50 μA (Section 9.6) for the phase-comparator current setting, the phase-comparator gain is:

$$\frac{50\ \mu A \times 1000}{6.28}$$

These values can be inserted into the t_1, t_2, and t_3 equations to find the values for C_1, C_2, and R_2, as described in the following example. Calculate the values for a third-order loop filter with the following parameters:

Frequency to be synthesized	800 MHz
Reference frequency	100 kHz
Division ratio, N	$\dfrac{800\text{ MHz}}{100\text{ kHz}} = 8000$
Natural loop frequency, n	500 Hz
VCO gain factor, K_2	6.28×10 MHz/volt
Phase margin, θo	45°
Phase-comparator current setting	50 μA (Section 9.6, $G_1/G_2 = 0$)

Assuming that $R_1 = 1$ kΩ, then the equivalent phase-comparator gain (K_1, in V/radian) is:

$$\frac{50\ \mu A \times 1000}{6.28} = 0.00796\ \text{V/radian}$$

$$t_3 = \frac{-\tan 45° + \dfrac{1}{\cos 45°}}{6.28 \times 500} = \frac{0.4142}{3140}$$

$$t_3 = 1.319 \times 10^{-1}$$

$$t_2 = \frac{1}{(6.28 \times 500)^2 \times 1.319 \times 10^{-4}} = 7.69 \times 10^{-4}$$

$$t_1 = \frac{7.96 \times 10^{-3} \times 6.28 \times 10\ \text{MHz/V}}{} \left| A \right|^{1/2}$$

where $A = \dfrac{1 + (6.28 \times 500)^2\,(7.69 \times 10^{-4})^2}{1 + (6.28 \times 500)^2 \times (1.319 \times 10^{-4})^2}$

$$t_1 = \frac{499{,}888}{7.89 \times 10^{10}} \left| \frac{6.832}{1.1714} \right|^{1/2}$$
$$t_1 = 6.33 \times 10^{-6} \times 2.415$$
$$t_1 = 1.53 \times 10^{-5}$$

Because $t_1 = C_1 R_1$, C_1 is:

$$\frac{1.53 \times 10^{-5}}{1000} = 0.0153 \ \mu F$$

For Fig. 9.11A:
$$t_2 = R_2 (C_1 + C_2)$$

For Fig. 9.11B:
$$t_3 = C_2 R_2$$

Substituting for C_2:

$$t_2 = R_2 \left| C_1 + \frac{t_3}{R_2} \right|$$

$$t_2 = R_2 C_1 + t_3$$

$$R_3 = \frac{t_2 - t_3}{C_1} = \frac{(7.69 \times 10^{-4}) - (1.319^{-4})}{0.0153 \times 10^{-6}}$$

$$R_2 = 41640 = 4.164 \ k\Omega$$

$$t_3 = C_2 R_2$$

$$C_2 = \frac{t_3}{R_2} = \frac{1.319 \times 10^{-4}}{41640} = 3.17 \ nF$$

for Fig. 9.11B:

$$t_1 = C_1 R_1$$

$$C_1 = \frac{1.53 \times 10^{-5}}{1000} = 0.0153 \ \mu F$$

$$t_2 = C_1 R_2$$

$$R_2 = \frac{7.69 \times 10^{-4}}{0.0153 \times 10^{-6}} = 50.261 \ k\Omega$$

$$t_3 = C_2 R_3$$

Because both values are independent of the other components, either C_2 or R_3 can be chosen, and the other calculated.

Assume that $R_3 = 1 \ k\Omega$.

$$C_2 = \frac{1.319 \times 10^{-4}}{1000} = 1.319 \times 10^{-7} = 0.1319 \ \mu F$$

Chapter
10

Optimizing High-Speed Frequency Dividers

This chapter is devoted to optimizing high-speed frequency-divider ICs. Although such devices are digital, they must be considered as RF when optimizing layout, coupling between parallel traces, impedance matching, etc. Before getting into specific divider ICs, review the many general problems associated with high-speed dividers.

10.1 PC Boards for High-Speed Dividers

Both the electrical performance and the mechanical/thermal performance must be considered when selecting PC boards for high-speed divider circuits. For example, using a $1/16''$ fiberglass PC board might be desirable mechanically, but a 50-Ω stripline on this thickness of board is about $5/32''$ wide, and thus too wide to pass between the pins of an IC.

Most of the heat conducted from a dual-in-line IC package is removed from the bottom of the package. Less than 10% is conducted out by the leads. Because of the cavity between the chip and lid, relatively little heat is passed through the top of the package. For this reason, a double-layer PC board is recommended, and the ground plane should be on the top surface. Where $1/32''$ board material is used, a top-surface ground plane will add substantially to the heat-dissipation capabilities of the board.

10.2 Components Used at High Frequencies

As in the case of any high-speed IC (not just dividers), the choice of external components requires additional care. For example, carbon-composition resistors are more nearly resistive at high frequencies than either carbon or metal-film types. Carbon-composition resistors are also readily available in very small physical sizes. Bypass capacitors must be chosen carefully if they are to act as low impedances. This is because the leads produce an increasing impedance as the frequency increases above the series resonant frequency of the capacitor.

As a guide, a 1000-pF disc ceramic capacitor with ¼-inch leads will be self-resonant at about 75 MHz, and will appear as an inductive impedance of about 22 Ω at 800 MHz. The use of chip capacitors is recommended at frequencies above 500 MHz, although ceramic capacitors with short leads are often acceptable.

10.3 Single-Point Grounding

Figure 10.1 shows how single-point grounding can be used in high-speed (or high-frequency) circuits. All of the bypass capacitors are returned to a single point, and this point is returned to the ground plane. Also, the output load resistors have their grounded ends connected together, and a common return is used. Because the currents in the resistors are in antiphase, the inductive effects are canceled, and the path followed by the relatively large output currents is controlled. Defining the ground-current path is more important in applications like frequency synthesis (Chapter 9), where a relatively large part of the system might be on one PC board.

10.4 Impedance Matching PC Board Traces

When PC board traces are terminated in impedances that are substantially different from the trace impedance, there are voltage variations along the trace (standing waves, as described in Section 2.18.1). If the trace can be terminated by an impedance near to that of the trace impedance, the voltage variations will be at a minimum.

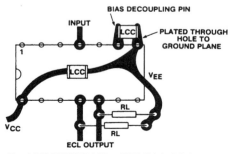

Figure 10.1 Single-point grounding for high-speed circuits (*GEC Plessey Semiconductors,* Professional Products IC Handbook, *p. 4-55*).

The impedance of PC-board traces depends on board thickness, trace width, and the board material. Assuming a $1/16''$ glass-fiber epoxy board, the approximate impedances are: 100 Ω for a 1-mm trace width, 75 Ω for 2 mm, and 50 Ω for 4 mm. With these impedances, it is assumed that the ground plane is on the opposite side of the board, and that it covers the area completely.

10.5 Frequency Range and Input Level

Figure 10.2 shows an example of a *guaranteed operating window* for a high-speed divider. Both the input signal level and the operating frequency are involved. For example, excessive input can cause permanent damage to the IC. Even if there is no damage, an excessive input signal can cause the divider to miscount, especially when cold. Running the divider IC at too low a level can also cause problems—even though the level is within the "typical performance" specification of the device.

Figure 10.3 shows the effect of coupling between traces. Here, an ECL output signal on pin 6 of the IC couples 60 mV of signal to the input at pin 3 (at a frequency of 500 MHz). Such a level of coupling can

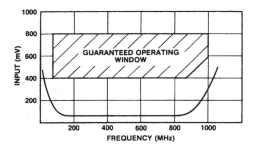

Figure 10.2 Example of guaranteed operating window curve (*GEC Plessey Semiconductors,* Professional Products IC Handbook, *p. 4-55*).

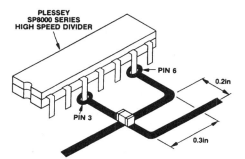

Figure 10.3 Effects of coupling between parallel traces (*GEC Plessey Semiconductors,* Professional Products IC Handbook, *p. 4-56*).

lead to divider jitter if the input signal is low. As a general precaution, try to keep all inputs and outputs well separated when higher frequencies are involved. This includes lines to modulus-control pins on two-modulus dividers (especially ECL dividers).

Typically, with the frequency range specified for high-speed dividers, it is assumed that a sine-wave input will be used, and that the dividers are edge triggered. In many cases, dividers will operate at frequencies below those specified in the data-sheet operating window, provided that the input signal has a suitable high slew rate (typically 100 to 200 volts/microsecond). The increase in slew rate should be accomplished by shaping the input signal, rather than by simply overdriving the IC.

10.6 Dividers with ECL Outputs

High-speed dividers with ECL outputs sometimes require external load resistors. Always consult the datasheet. If required, be certain that the resistor leads are as short as possible, and that noninductive resistors are used. Where an ECL-output divider drives another divider, use ac coupling. This is because few dividers are strictly ECL compatible at their inputs.

10.7 Dividers with Open-Collector TTL Outputs

Open-collector TTL outputs are relatively slow. Although the negative edge is limited in speed by the turn-on time of the output transistor, the rising edge is limited by the external load resistor and capacitance to ground. In practice, this means that short, narrow traces are required for the following device, and a minimum fan-in load must be provided. Also, open-collector TTL outputs should not be used above about 10 MHz.

10.8 Dividers with True TTL Outputs

Dividers with true TTL outputs are not as limited as those with open-collector TTL outputs because of the active pull-up. However, the use of true TTL outputs at frequencies above 25 to 30 MHz is not recommended—especially into capacitive loads. Loads of more than 30 pF should not be driven faster than about 15 MHz. Notice that the current drawn by true TTL outputs increases with increasing load capacitance.

10.9 Dividers with CMOS Outputs

Most CMOS outputs are TTL compatible. However, CMOS outputs are not guaranteed to meet TTL levels at TTL currents. As a result, it is generally better not to use CMOS-output devices to drive TTL loads directly. When an interface is required, use an active transistor interface between a CMOS output and a TTL input.

10.10 ECL-TTL Interface

Figure 10.4 shows an ECL-TTL interface circuit that uses a line receiver. Simple circuits with one or two transistors cannot be guaranteed to work over all of the ECL output voltages and temperature ranges.

10.11 Interfacing High-Speed Dividers

The following is a summary of rules for interfacing high-speed dividers.

1. Treat all high-speed divider ICs as RF devices. Use conventional high-frequency RF layout and component-selection guidelines. For example, 0.1-μF ceramic capacitors do not provide a good bypass at 1.5 GHz.
2. Do not use CMOS outputs to drive TTLs directly.
3. Do not use open-collector outputs above 10 MHz.
4. Observe the input requirements of high-speed dividers. For example, always check the guaranteed input operating area and temperature ranges.

10.12 Matching the Input Impedance of High-Speed Dividers

The simplest method to determine the input impedance of a high-speed divider is to use the Smith chart, as described in Section 2.10. That is why the datasheets for dividers usually include a Smith-chart presentation for the input impedance.

No matter what method is used to find impedance (and the corresponding component values that match that impedance), remember

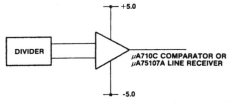

Figure 10.4 ECL-TTL interface circuit (*GEC Plessey Semiconductors,* Professional Products IC Handbook, *p. 4-56*).

that the calculated impedance is correct for only one frequency. It is difficult, at best, to provide wideband impedance matching. For that reason, matching networks should be low Q (Chapter 2). The obvious reason for this is that high-Q networks are narrowband, so any mismatch results in severe loss of signal. Another reason for using low Q is that low-Q matching networks are far more tolerant of variations in component values. As a guideline, when Q is doubled, the power loss is increased by a factor of four.

10.13 Optimizing Variable-Modulus Dividers

This section is devoted to optimizing variable-modulus (or dual-modulus) dividers. The GEC Plessey SP8685 is selected as an example. Figures 10.5, 10.6, 10.7, and 10.8 show the functional diagram, electrical characteristics, guaranteed operating window, and timing diagram, respectively. The SP8685 is an ECL variable-modulus divider, with ECL 10-kΩ compatible outputs. The IC divides by 10 when either of the ECL control inputs $\overline{PE1}$ or $\overline{PE2}$ is in the high state, and by 11 when both are low (or open circuit), as follows:

$\overline{PE1}$	$\overline{PE2}$	Division ratio
L	L	11
H	L	10
L	H	10
H	H	10

The input impedance of the IC depends on the operating frequency, as shown in the Smith chart of Fig. 10.9. The chart was prepared under the following test conditions: supply voltage = -5.2 V, ambient temperature = 25°C, frequencies in MHz, and impedances normalized to 50 Ω. Figure 10.10 shows the test circuit for both the Smith chart and the electrical characteristics of Fig. 10.6.

As shown by the typical application circuit of Fig. 10.11, the clock input is biased internally and is coupled to the signal source with a suitable capacitor. The input signal path is completed by an input-reference decoupling capacitor, which is connected to ground.

If no signal is present, the IC will self-oscillate. If this is undesirable, self-oscillation can be prevented by connecting a 15-kΩ resistor from clock input (pin 12) to V_{EE}. The resistance will reduce the input sensitivity by about 100 mV.

The IC will operate at frequencies down to dc. However, the slew rate of the input signal must be greater than 100 V/μs.

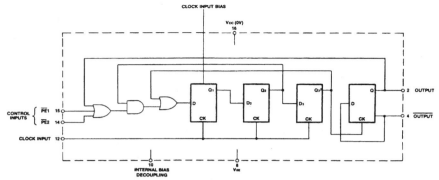

Figure 10.5 SP8685 functional diagram (*GEC Plessey Semiconductors*, Professional Products IC Handbook, *p. 3-21*).

ELECTRICAL CHARACTERISTICS

Supply Voltage: V_{CC} = 0V V_{EE} = -5.2V ± 0.25V
Temperature: A Grade T_{amb} = -55°C to +125°C
B Grade T_{amb} = -30°C to +70°C

Characteristic	Symbol	Min.	Max.	Units	Conditions	Notes
Maximum frequency (sinewave input)	f_{max}	500		MHz	Input = 400-800mV p-p	
Minimum frequency (sinewave input)	f_{min}		50	MHz	Input = 400-800mV p-p	Note 6
Power supply current	I_{EE}		70	mA	V_{EE} = -5.2V	Note 6
Output high voltage	V_{OH}	-0.87	-0.7	V	V_{EE} = -5.2V (25°C)	
Output low voltage	V_{OL}	-1.8	-1.5	V	V_{EE} = -5.2V (25°C)	
$\overline{PE}$ input high voltage	V_{INH}	-0.93		V	V_{EE} = -5.2V (25°C)	
$\overline{PE}$ input low voltage	V_{INL}		-1.62	V	V_{EE} = -5.2V (25°C)	
Clock to output delay	t_D		6	ns		Note 7
Set-up time	t_s	2		ns		Note 7
Release time	t_r	2		ns		Note 7

NOTES
1. Unless otherwise stated, the electrical characteristics shown above are guaranteed over specified supply, frequency and temperature range.
2. The temperature coefficient of V_{OH} = +1.63mV/°C, V_{OL} = +0.94mV/°C and of V_{IN} = +1.22mV/°C but these are not tested.
3. The test configuration for dynamic testing is shown in Fig.6.
4. The set up time t_s is defined as minimum time that can elapse between L → H transition of control input and the next L → H clock pulse transition to ensure that +10 is obtained.
5. The release time t_r is defined as the minimum time that can elapse between H → L transition of the control input and the next L → H clock pulse transition to ensure that the +11 mode is obtained.
6. Tested at 25°C only.
7. Guaranteed but not tested.

Figure 10.6 SP8685 electrical characteristics (*GEC Plessey Semiconductors*, Professional Products IC Handbook, *p. 3-22*).

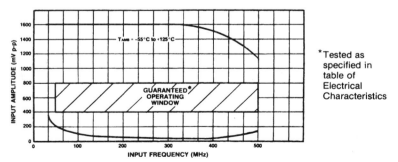

*Tested as specified in table of Electrical Characteristics

Figure 10.7 SP8685 guaranteed operating window (*GEC Plessey Semiconductors*, Professional Products IC Handbook, *p. 3-22*).

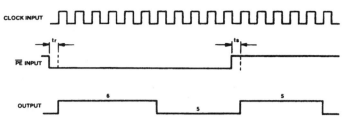

Figure 10.8 SP8685 timing diagram (*GEC Plessey Semiconductors, Professional Products IC Handbook, p. 3-22*).

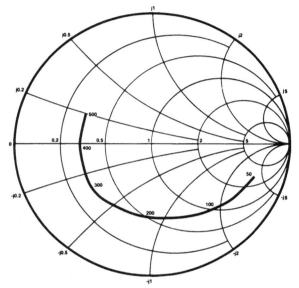

Figure 10.9 SP8685 input impedance Smith chart (*GEC Plessey Semiconductors,* Professional Products IC Handbook, *p. 3-23*).

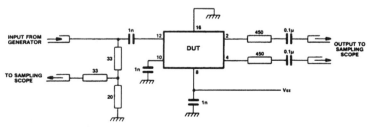

Figure 10.10 SP8685 test circuit (*GEC Plessey Semiconductors, Professional Products IC Handbook, p. 3-23*).

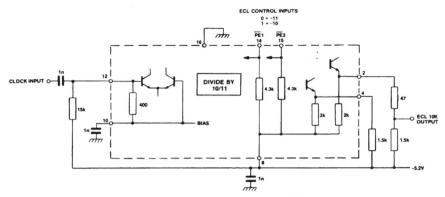

Figure 10.11 SP8685 typical application circuit (*GEC Plessey Semiconductors, Professional Products IC Handbook, p. 3-24*).

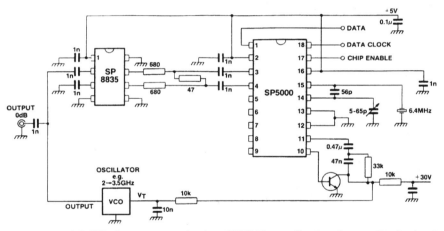

Figure 10.12 3.3-GHz frequency synthesizer (*GEC Plessey Semiconductors,* Professional Products IC Handbook, *p. 4-64*).

The two outputs are compatible with ECL II, but can be interfaced to ECL 10K, as shown in Fig. 10.11.

The $\overline{PE}$ inputs are ECL III/10K compatible, and include 4.3-kΩ internal pulldown resistors. As a result, unused $\overline{PE}$ inputs can be left open.

All external components must be suitable for the selected operating frequency. Use RF layout procedures (ground planes, single-point grounding, short leads for external components, etc.) for all high-speed divider circuits.

10.14 Using Dividers in Microwave Synthesizers

The high-speed dividers described in this chapter are used primarily in frequency synthesizer (FS) circuits, either PLLs or FLLs (frequency-locked loops). This section is devoted to using dividers in FS circuits that operate in the microwave (3.5 GHz) region.

As covered in Section 1.6, FS circuits include a VCO that is controlled by a feedback network so that the VCO oscillates at a predetermined frequency. In most cases, the synthesized frequency can be altered (by external programming) to tune the oscillator over a range of frequencies (a band). The divider (or prescaler) is used in such a system to reduce the VCO frequency to one that can be compared with a standard or known reference frequency (usually obtained from a crystal-controlled oscillator and divider). This frequency comparison can be made in several ways. For example, the most common way is to make a phase comparison using a PLL. Another method is to make a frequency comparison using a frequency-counting microprocessor in a frequency-locked loop.

Figure 10.12 shows a 3.5-GHz FS circuit using an SP8835 divider and an SP5000 synthesizer. A similar synthesizer is described in Chapter 11. The only interface required between the SP8835 and the SP5000 is an attenuator that prevents the SP8835 from overloading the SP5000 input. The rest of the application circuit is identical to the standard application circuits of the two devices, as shown in their respective data sheets. Using the values shown and a 4-MHz crystal in the reference oscillator, the circuit can be programmed to synthesize frequencies up to 3.5 GHz, with a minimum frequency-step size of 500 kHz. The range of frequencies over which the synthesizer will operate is determined by the VCO used.

10.15 Using Dividers in VHF Synthesizers

Figure 10.13 shows a serially programmable VHF synthesizer using an SL562 low-noise op amp, an SP8793 dual-modulus divider (or prescaler), and an NJ8822 synthesizer. A similar synthesizer is covered in Chapter 11. The NJ8822 is programmed by a serial microprocessor interface.

The VCO is a JFET (J310) oscillator using a transmission line (105 mm of solid 75-Ω coax) as the resonator. This eliminates the need for a coil. The VCO is modulated by applying the audio to the cathode of a reverse-biased PIN diode (BA182).

The loop filter uses the SL562, which, with the values shown, has a loop bandwidth of 60 Hz and a damping factor of 0.6. The loop filter is followed by a low-pass filter pole at 3.7 kHz to attenuate the 12.5-kHz reference sidebands.

Optimizing High-Speed Frequency Dividers

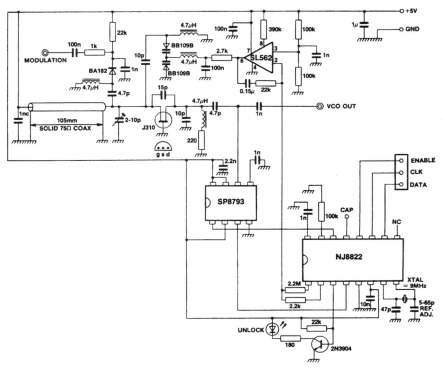

Figure 10.13 Serially programmable VHF synthesizer (*GEC Plessey Semiconductors, Professional Products IC Handbook, p. 4-46*).

The lock-up time for a 1-MHz change in frequency is 80 ms. The output frequency range is 144 to 146 MHz, with a level of +3 dBm into 50 Ω. The circuit operates with a supply voltage of 5 V ±0.5 V, with a temperature range of −30°C to +70°C. A frequency drift of 3 kHz might occur between these two temperature extremes. This is because of the uncompensated reference oscillator.

Chapter 11

Optimizing Frequency Synthesizer Design

This chapter is devoted to optimizing the design of a complete FS circuit using an off-the-shelf IC. The GEC Plessey NJ8820/21/23 is selected as an example. These ICs are versatile CMOS FS controllers. The differences among the three devices are in the hardware programming methods.

11.1 Basic Single-Loop FS PLL

Figure 11.1 shows the basic single-loop PLL circuit using the NJ8820/21/23. A two-modulus prescaler is used to divide the VCO frequency down to a suitable range for use in the FS IC. The NJ8820/21/23 is programmed by 8 of 4-bit words on the data inputs. The addresses for these words can be obtained internally or externally, and they appear on the data-select inputs/outputs. To program any frequency, it is necessary to program the A counter, the M counter, and the reference (R) counter. These counters are 7, 10, and 11 bits long, respectively.

11.2 Addressing in the Self-Programming Internal Mode

In this mode, the reference-oscillator (either an internal crystal oscillator or from an external source) signal is divided in the reference counter by 64 and a DATA READ cycle commences every $1024/f_{osc}$ seconds.

In the DATA READ cycle, the MEMORY ENABLE pin is pulled low, and the DATA SELECT outputs, DS0, DS1, and DS2, count in binary from 0 to 7. This provides addresses for the DATA on D0, D1, D2, and D3. The data bits are transferred to internal latches during the trailing edge of the DATA SELECT pulses, as shown in Fig. 11.2.

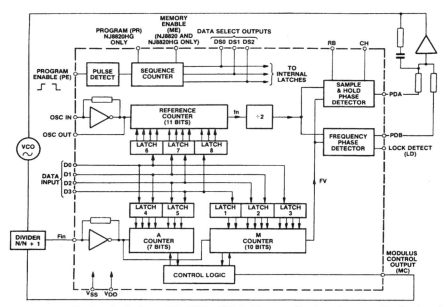

Figure 11.1 Basic single-loop FS PLL (*GEC Plessey Semiconductors,* Professional Products IC Handbook, *p. 4-40*).

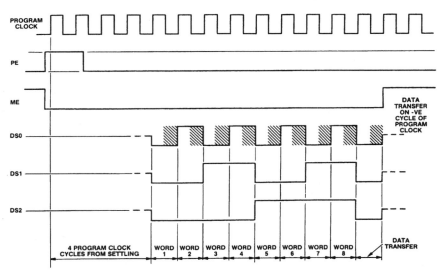

Figure 11.2 PLL data-selection timing diagram (*GEC Plessey Semiconductors,* Professional Products IC Handbook, *p. 4-41*).

The program clock is internally derived and is at a frequency of $f_{osc}/64$. The PE (program enable) pin is grounded, and the cycle continuously repeats. The self-programming mode is not generally recommended because noise might be picked up by the PLL.

11.3 Addressing in the Single-Shot Internal Mode

In this mode, the PE pin is provided with a pulse input. The pulse initiates a data-read cycle, as described in Section 11.2. At the end of the cycle, the ME (memory enable, NJ8820 and NJ8820HG only) pin goes high and system power consumption is minimized. A power-on initiation is required in this mode (where the application of power to the IC is sensed and the programming cycle is started). To avoid corruption of the data, a delay of 53,248 cycles of the reference-oscillator frequency is provided before the programming cycle begins. This delay is approximately 5 ms for a 10-MHz reference frequency.

11.4 Addressing in the External Mode

In this mode (which is the preferred mode), the address is presented to DS0, DS1, and DS2, and a pulse is applied to the PE pin to transfer data to the internal latches. The data bits are transferred from the latches to the counters simultaneously with the transfer of data into Latch 1 (thus, this word should be the last one entered).

11.5 Frequency Synthesizer IC Pin Functions

Before optimizing the design, review the IC pin functions. Although the following descriptions apply specifically to the pins of the NJ8821, the descriptions are similar for the NJ8820 and NJ8823.

The PDA pin is the analog output from the sample-and-hold phase comparator for use as a *fine error signal*. The output is $V_{DD} - V_{SS}/2$ when the loop is locked. The PDA voltage increases when the FV phase lead increases, and decreases when the FR phase lead increases. The PDA output is linear over only a narrow phase window. The width of the window is programmed by a voltage at the R_B pin.

The PDB pin is a three-state output from the phase/frequency detector for use as a *coarse error signal*. If *FV* is greater than *FR*, or with *FV* leading, the output is a series of positive pulses. If *FV* is less than *FR*, or with *FR* leading, the output pulses are negative. If *FV* = *FR* and the phase error is within the PDA window, the output is a high impedance.

The LD pin is an open-drain lock-detect output. The LD output is low when the phase error is within the PDA window (in lock). The LD output is high impedance at all other times.

The FIN pin is the input to the main counters. When the pin is driven from a prescaler, the input signal can be ac coupled. If the input has a full logic swing, the FIN pin can be dc coupled.

The V_{SS} pin is the negative supply, and is normally connected to ground.

The V_{DD} pin is the positive supply (typically 5 V).

The OSC.IN and OSC.OUT pins form an on-chip reference oscillator when a parallel-resonant crystal is connected across them. Capacitors of an appropriate value are also required between each end of the crystal and ground to provide the necessary additional phase shift (to sustain oscillation). An external crystal-generated reference signal can be applied to OSC.IN (instead of a crystal across both pins). Such an external crystal-controlled input can be a low-level signal that is ac coupled into OSC.IN. If the external signal has a full logic swing, the input can be dc coupled into OSC.IN. The program range of the reference counter is 6 to 4094, in steps of 2, with the division ratio being twice the programmed number.

Information on the D0 through D3 pins is transferred to the internal latches during the appropriate data-read time slot (D3 on the MSB and D0 on the LSB).

The PE pin is used as a strobe for the data. A logic high on the PE pin transfers data from the data pins to the internal latch selected by the address (data select) lines. A logic zero on the PE pin disables the data lines.

The DS0 through DS2 pins are the data-select inputs used to control the addressing of data latches.

The MC pin provides a signal to control an external dual-modulus prescaler. The modulus-control level is low at the beginning of a count cycle; it remains low until the A counter completes its cycle. The modulus control then goes high and remains high until the M counter completes its cycle. At this point, both counters are reset. This gives a total division ratio of $M.N + A$, where N and $N + 1$ represent the dual-modulus prescale values.

The program range of the A counter is 0 to 127; therefore, it can control prescalers with a division ratio up to and including divide by 128/129.

The program range of the M counter is 3 to 1023. For correct program operation, M must be equal to (or greater than) A. Where every possible channel is required, the minimum division ratio should be $N^2 - N$.

The R_B pin provides a means to control the sample-and-hold (S/H) phase detector. A gain-programming resistor should be connected between the R_B pin and V_{SS}.

The CH pin also provides for control of the S/H phase detector. An external capacitor should be connected between CH and V_{SS} to set the desired hold time of each sample.

11.6 Programming Considerations

Timing for the IC is generated externally, normally from a microprocessor. This allows the user to change the data in selected latches. The following data map occurs when the PE pin is used as a strobe for the data.

WORD	DS2	DS1	DS0	D3	D2	D1	D0
1	0	0	0	M1	M0	—	—
2	0	0	1	M5	M4	M3	M2
3	0	1	0	M9	M8	M7	M6
4	0	1	1	A3	A2	A1	A0
5	1	0	0	—	A6	A5	A4
6	1	0	1	R3	R2	R1	R0
7	1	1	0	R7	R6	R5	R4
8	1	1	1	—	R10	R9	R8

When the PE pin is high, data bits are transferred into the selected latch. With the PE pin low, the data pins are disabled and the data bits are retained on the selected latch. Data transfer for all internal latches into the counters occurs simultaneously with the transfer of data into latch 1. Therefore, this would normally be the final latch addressed during each channel change. Figure 11.3 shows the timing sequence.

When reprogramming, a reset-to-zero state is followed by reloading with the new counter values. This means that the synthesizer loop lock-up time is well defined and less than 10 ms. (If shorter lock-up times are required, when making only small changes in frequency, the nonresettable version, NJ8823, should be considered.)

11.7 Phase Comparators

A standard digital phase/frequency detector driving a three-state output (PDB) provides a coarse error signal to enable fast switching between channels. This output is active until the phase error is within the S/H phase-detector window, at which point the PDB output becomes high impedance. The phase lock is indicated at this point with a low level on LD. The S/H phase detector provides a fine error signal to provide further phase adjustment, and to hold the loop in lock.

An internally generated ramp controlled by the digital output from both the reference and main divider chains is sampled at the

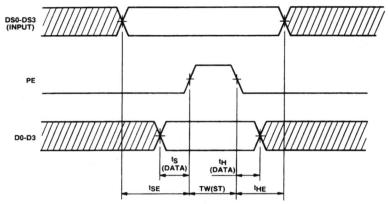

Figure 11.3 PLL data-transfer timing (*GEC Plessey Semiconductors, Professional Products IC Handbook, p. 2-45*).

reference frequency to provide the fine error signal PDA. When in phase lock, PDA is typically at $V_{DD} - V_{SS}/2$. Any offset from this is proportional to phase error.

The relationship between the offset voltage and the phase comparator is the phase-comparator gain, which is programmed with the external resistor at R_B. An internal 50-pF capacitor is used in the S/H comparator. The value of the R_B resistor should be chosen to provide the required gain at the reference frequency used. Figure 11.4 shows the relationship of R_B, gain, and reference frequency.

As an example, to get a gain of 380 V per radian at 10 kHz requires an R_B of about 39 kΩ. The value of the hold capacitor at the CH pin is not crucial. A typical value is 470 pF. A smaller value can be used if the sideband performance is not crucial. Figure 11.4 shows the gain normalized to a 1-Hz comparison frequency. To get the value for any other frequency, divide the value of the gain-frequency product by the desired frequency.

11.8 Reference Divider Programming

The reference divider produces the comparison frequency required by the synthesizer, and is programmable from 6 to 4094 in steps of 2. The division ratio is twice the programmed number. For example, assuming that a 10-MHz crystal is used, and that a 12.5-kHz reference is required, the counter ratio would be programmed to provide a ratio of 100,000/12.5 = 800. The actual programming would then be 400. Because the reference counter is an 11-bit binary device, the 400 programming would be entered as 00110010000.

11.9 A and M Divider Programming

The A counter is a 7-bit binary counter and the M counter is a 10-bit binary counter. The basic programming calculations are as follows:

1. The A counter should contain x bits, such that $2^x = M$.
2. If more bits are included in the A counter, these should be programmed to zero. For example:

$$M = 64 = 6 \text{ bits}$$
$$A = 10 \text{ bits}$$

then the 4 MSBs are programmed to zero.

3. The M and A counters are treated as being combined so that the MSB of the M counter is the MSB of the total, and the LSB of the A counter is the LSB of the total. For example: A synthesizer operating from 430 to 440 MHz in 25-kHz steps uses a 64/65 divider, and the control circuit uses binary counters.

$P = f/f_{ref}$ and $f_{ref} = channel\ spacing = 25$ kHz
$P_{min} = 430/0.025 = 17200$
$P_{max} = 440/0.025 = 17600$

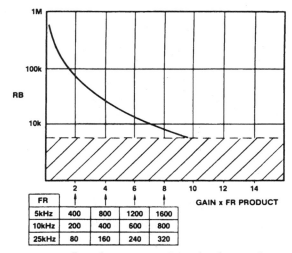

Figure 11.4 R_B value versus gain and reference frequency (*GEC^B Plessey Semiconductors*, Professional Products IC Handbook, *p. 2-46*).

The minimum possible divide ratio is $N^2 - N = 4032$, where N is the two-modulus divider ratio.
The *maximum allowable loop delay* $= 64/(440 \times 10^6) = 145$ ns. The total divide ratio (P) is given by:

$$P = NM + A$$

$N = 64$ (because a 64/65 divider is used), P_{min} is 17200. Therefore, $17200 = 64M + A$, and M is equal to or greater than A. Let $A = 0$. Then, $M_{min} = 17200/64 = 268.75$ rounded off to 268, and $M_{max} = 17600/64 = 275$.

Thus, the M counter must be programmable from 268 to 275 as required. The M counter must have at least 9 bits.

As an example, for a frequency of 433.975 MHz:

$$P = 433.97/0.025 = 17359$$

and

$$M = 17359/64 = 271.2343$$

The A counter is programmed for the remainder:

$$0.2343 \times 64 = 15$$

From this rather tedious calculation, the A counter is programmed to 15 and the M counter to 271. The output frequency can now be checked:

$$P = NM + A$$
$$= (271 \times 64) + 15 = 17359$$

and this is the required divider ratio, in decimal. Remember that the A and M counters are binary dividers and that the decimal values must be converted, as is the case for the reference divider (Section 11.8).

11.10 Calculator Program for Synthesizer Programming

The programming calculations described in Section 11.9 can be made on most Hewlett Packard calculators using the following program:

Line	Function	Display
001	hLBLA	25 13 11
002	ENTER	31
003	RCL0	24 0
004	—	71
005	ST02	23 2
006	RCL1	24 1
007	—	71
008	ST03	23 3
009	hFRAC	25 33
010	ENTER	31
011	RCL1	24 1
012	X	61
013	ST04	23 4
014	RCL3	24 3
015	ENTER	31
016	RCL3	24 3
017	hFRAC	25 33
018	—	41
019	ST03	23 3
020	hPSE	25 74
021	hPSE	25 74
022	RCL4	25 4
023	hRTN	25 12

To use the calculator program, enter the comparison frequency in STO0, and the dual-modulus prescaler ratio in STO1 (this is the value of N in an $N/N + 1$ divider).

Enter the frequency to be synthesized in Hz and press the R/S button. The calculator will flash twice and display the decimal value of M. Pressing R/S again will display the value for the A counter. The M-counter value is in STO3. The A-counter value is in STO4.

The values of M, A, and R must be fed into the NJ8820/21 for each value of frequency required. (In this example, the value of R is assumed to be constant.) In each case, the LSB is identified by the heading M_0, A_0, or R_0.

The NJ8820 and NJ8821/23 require 32 bits of data to be transferred for each value of frequency. These 32 bits are composed of the 28 bits from the R, M, and A counters (11 + 10 + 7), plus 4 redundant bits. The NJ8820 receives data from a PROM. The NJ8821/23 receive data from a microprocessor. The steps described in this chapter are a summary of the design procedures for a complete frequency synthesizer. It is recommended that the datasheets and any available application notes be followed in an actual design situation.

Chapter 12

Optimizing Direct Frequency Synthesizers

This chapter is devoted to optimizing direct frequency synthesizer (DFS) circuits. The frequency-synthesizer circuits covered thus far use some form of PLL to produce signals at a specific frequency, or across a band of frequencies. DFS circuits are available in IC form. The GEC Plessey SP2001 is selected as an example. Before getting into the IC characteristics, review the DFS principles and some common circuit problems.

12.1 The Basic DFS Circuit

Figure 12.1 shows the basic elements of a DFS circuit. Note that the term *DFS* can also mean *digital frequency synthesizer* because direct frequency synthesis is a digital process, involving accumulators, adders, latches, ROMs, and DACs. In the basic circuit, the accumulator provides an address to the ROM in which the desired wave shape is stored. The output of the ROM is fed to the DAC to provide the analog output waveform.

The address to the ROM is produced by adding a binary number (K) to the content of the N-bit accumulator. The resulting sum is transferred to the latch at each reference clock cycle. Each time the adder overflows, a new cycle is started, and the rate at which this occurs is the output frequency (which is set by the programmed number, K).

The output frequency is determined by the numbers N and K (the number of bits in the accumulator and the programmed number).

$$f_{\text{out}} = \frac{Kf_c}{2^N}$$

where f_c is the clock frequency.

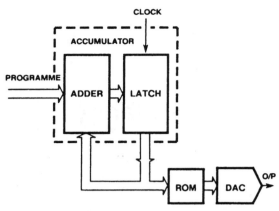

Figure 12.1 Basic elements of a DFS circuit (*GEC Plessey Semiconductors*, Professional Products IC Handbook, p. 4-66).

The minimum output frequency is produced when $K = 1$, or:

$$f_{\text{out}} = \frac{f_c}{2^N}$$

The smallest increase is from $K = 1$ to $K = 2$, or:

$$\frac{f_c}{2^N} \text{ to } \frac{2f_c}{2^N} \text{ and thus } \frac{f_c}{2^N}$$

The step size is fixed by N and so, for a 16-bit accumulator, is:

$$\frac{f_c}{65536}$$

The internal ROM is programmed with a half-cosine wave so that the amount of ROM required is at a minimum (128×8). The ROM output is an 8-bit word used to drive the DAC.

12.2 SP2001 Circuit Description

Figures 12.2, 12.3, and 12.4 show the block diagram, typical application circuit, and electrical characteristics of the SP2001. The IC is an ECL 100K-compatible device that generates a DAC code required for an output sine wave at any frequency up to 100 MHz. The SP2001 has a 16-bit input bus, providing a step size and minimum output frequency of 4 kHz (using a 262.144-MHz clock). Both Figs. 12.2 and 12.3 show the SP2001 used with an SP98608 DAC.

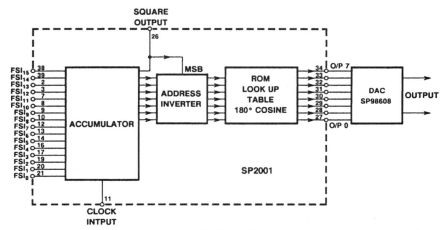

Figure 12.2 SP2001 block diagram (*GEC Plessey Semiconductors*, Professional Products IC Handbook, *p. 2-33*).

The IC needs no reactive components, except power-supply decoupling capacitors, and it is under full digital control at all times. Frequency accuracy is set by an external oscillator. Phase noise on the synthesizer output is dominated by the clock performance. The fully digital system does not contain control loops (as does a PLL), so the "hop time" between discrete output frequencies is limited (in theory) only by the DAC settling time of about 5 ns.

In a practical circuit, a further delay of four clock periods is added to simplify the logic. The resultant delay of 20 ns (worst case) is about five times as fast as a typical PLL.

To avoid the need for a full 360 degrees in the ROM, the MSB output of the accumulator is used as a sine inverter, which, with the LSBs, forms a digitized triangular number sequence. The MSB of the accumulator provides a square-wave output through an ECL buffer, which can be used as a variable clock in digital systems. The 1-kΩ ROM, organized as 128 × 8 bits, contains the data for 180° stored in a cosine sequence. The cosine information is read twice for each cycle, thus providing 256 words of data in total. The data passes through retiming latches at each stage, including the output, to provide accurate data at the high clock range. Finally, the DAC reconstructs the output waveform.

A set input (pin 22) provides a "start from zero" function for test purposes. The set input sets all of the accumulator latches to zero so that the output of the ROM is in the "all ones" state. The frequency equation is:

$$f_{\text{out}} = \frac{f_{\text{clock}}}{2^{16} \times \text{input data}}$$

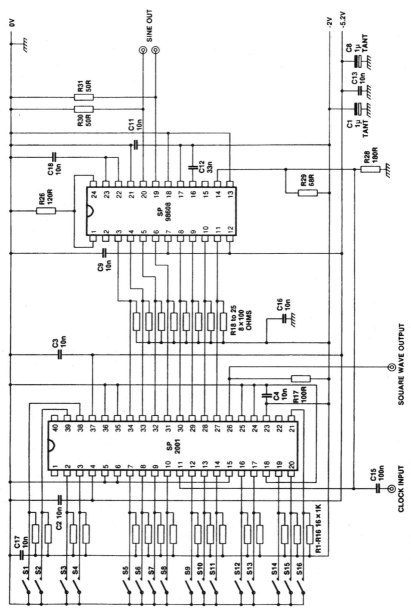

Figure 12.3 SP2001 typical application circuit (*GEC Plessey Semiconductors*, Professional Products IC Handbook, *p. 4-69*).

ELECTRICAL CHARACTERISTICS
Test conditions (unless otherwise stated)
$V_{EE} = -5.2V \pm 0.25V$, $V_{DD} = -2V \pm 0.1V$, $V_{CC} = GND$
A Grade $T_{amb} = -55°C$ to $+125°C$ (see note), B Grade $T_{amb} = -40°C$ to $+85°C$

Characteristic	Pin	Value			Units	Conditions
		Min	Typ	Max		
Supply current I_{EE}	4,37	220	290	370	mA	All inputs at -1.8V
Supply current I_{DD}	23	40	60	84	mA	Does not include ECL output current
Input HIGH voltage	FSI_{0-15},Clk	-1125		-880	mV	
Input LOW voltage	FSI_{0-15},Clk	-1810		-1520	mV	
Output HIGH voltage	26-34	-1125		-880	mV	Loaded with 100Ω to -2V
Output LOW voltage	26-34	-1810		-1520	mV	Loaded with 100Ω to -2V

Note: The SP2001 must be used with a suitable heatsink to maintain chip temperature below 175°C when operating at Tamb > 85°C for DG package and > 70°C for HG package. θ_{JA} DG = 36°C/W. θ_{JA} HG = 50°C/W.

ABSOLUTE MAXIMUM RATINGS
Storage temperature -65°C to 150°C
Max. junction temperature +125°C
Max voltage between V_{EE} and V_{CC} -7.0V to +0.5V
Input voltage (DC) V_{EE} to (V_{CC} + 0.5)V
Output current at $V_O = V_{OH}$ 20mA

Figure 12.4 SP2001 electrical characteristics (*GEC Plessey Semiconductors*, Professional Products IC Handbook, p. 2-34).

For example, for 5-kHz increments, f_{clock} = 327.68 MHz, and for 3.125-kHz increments, f_{clock} = 204.8 MHz.

As shown in Fig. 12.3, eight ECL data outputs from the SP2001 are connected directly to the DAC inputs with 100-Ω pull-down resistors. The pull-downs are essential because there are no internal loads for the ECL output transistors.

A square-wave output is available from pin 26, which uses an output stage identical to those feeding the DAC. The load resistance can be reduced to 50 Ω, if needed. In applications where the square-wave output is not required, it can be left as an open circuit to reduce power consumption. When only the square-wave output is required, the DAC-drive outputs can be left as an open circuit.

In the circuit of 12.3, the DAC is used in the Latched mode with the clock input feeding both the SP2001 and the DAC. The DAC can also be used in the Transparent mode by connecting pin 15 to ground and removing the clock. However, circuit performance will be degraded when compared to the Latched mode.

The clock inputs to the SP2001 and DAC are biased to the center of the ECL logic-swing range (about −1.4 V) using 68-Ω and 180-Ω resistors. This also provides a 50-Ω match to the clock input cable. The sine-wave clock signal is ac coupled to the biased clock inputs by C15. The optimum clock input level is +4 dBm, which equals the nominal 1-V p-p standard ECL logic swing.

The frequency-set data inputs to the SP2001 are ECL compatible with a nominal threshold voltage of about −1.4 V. ECL data bits can be connected directly to the pins, instead of the 16 switches and resistors shown. (These switches and resistors are part of a demonstration board.) Connections to CMOS input data are shown in Fig. 12.5. As shown, the connections are straightforward, provided that the CMOS logic (or the microprocessor providing the data inputs) is operating from the -5.2-V supply used by the SP2001.

TTL drive logic can also be used in essentially the same way as CMOS, provided that resistor pull-ups connected to the positive supply are used. This will ensure a logic-high level close to the positive supply.

12.3 Optimizing the SP2001 Layout

Figures 12.6, 12.7, and 12.8 show the SP2001 demonstration-board component layout, ground plane, and trace layout, respectively. These

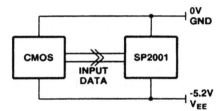

Figure 12.5 Interface to CMOS logic (*GEC Plessey Semiconductors,* Professional Products IC Handbook, *p. 2-35*).

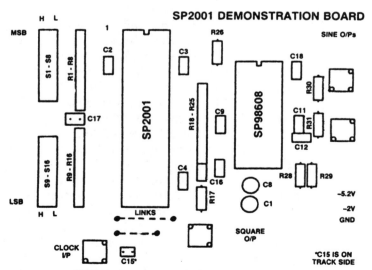

Figure 12.6 SP2001 demonstration board component layout (*GEC Plessey Semiconductors,* Professional Products IC Handbook, *p. 4-68*).

Optimizing Direct Frequency Synthesizers   191

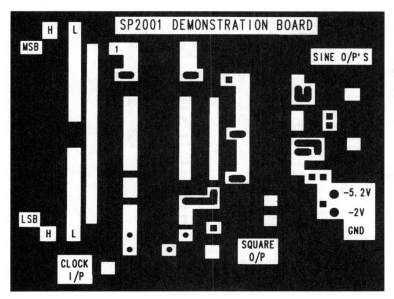

Figure 12.7 SP2001 demonstration board ground plane (*GEC Plessey Semiconductors,* Professional Products IC Handbook, *p. 4-68*).

illustrations also show the recommended methods for optimizing SP2001 performance. The board uses switches for programming. Using a 327.680-MHz clock, the switches provide the following binary selection:

LSB	provides 5-kHz steps
LSB + 1	provides 10-kHz steps
LSB + 2	provides 20-kHz steps

To program an output frequency, the program number is:

$$\frac{f_{\text{out}}}{5\text{ kHz}} \text{ expressed in binary}$$

The board requires supplies of -5.2 V and -2 V. The required clock input level is $+5$ dBm.

As always, care should be taken with the board layout. As shown in Figs. 12.6 through 12.8, the ground plane should be as continuous as possible, and common to both the SP2001 and the DAC. In the demonstration board, the ground plane serves as the V_{CC} supply point.

The $-5.2\ V_{\text{EE}}$ and -2 V supplies to both the SP2001 and DAC should be decoupled close to the device pins, preferably using surface-mounted

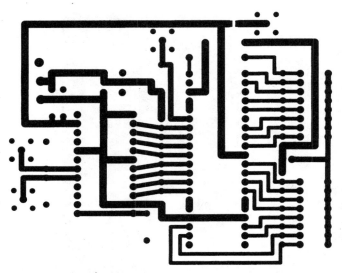

Figure 12.8 SP2001 demonstration board trace layout (*GEC Plessey Semiconductors*, Professional Products IC Handbook, *p. 4-67*).

10-nF capacitors. The eight ECL connections between the SP2001 and the DAC should be of equal length to ensure that the input data bits are presented without time skew.

The clock input to the SP2001 is fed through internal signal-conditioning circuitry to provide suitable internal ECL swings. The accumulator input is taken from the parallel ECL programming lines. When the switches are open, that particular data bit is programmed as a 1.

The output of the SP2001 ROM is fed to the SP68608 DAC (on the demonstration board). The DAC output is an approximate sine wave at the programmed frequency. The output level is about +5 dBm into 50 Ω. Because the output is a current source, the output impedance is determined by the value of the output resistance (50 Ω, in the case of the demonstration board).

Bypassing of the supply and DAC internal references are made by a combination of a chip ceramic and tantalum-bead capacitors. The pull-down resistors are single in-line resistor networks.

The value of resistor R26 sets the output level of the DAC. The approximate output voltage can be found by: $V\ p\text{-}p\ output = 128/R_{26}$. For a value of 120 Ω, as shown, the output voltage is approximately 1.1 V p-p.

12.4 Minimizing Spurious Outputs

To keep spurious outputs at a minimum, it is important that the DAC settle as quickly as possible. In this regard, a 10-bit DAC settling to

9 bits in a given time is better than a 12-bit DAC settling to 8 bits in the same time. A fast DAC is thus better than an identical slow DAC. (The recommended SP98608 8-bit DAC settles to $1/2$ LSB in 2 ns.)

Spurious outputs also depend on clock and output frequencies. A set of spurious phase-modulated sidebands exists at a carrier-to-sideband spacing dependent on the values of K, N, and f_c (Section 12.1). As a point of reference, measurements indicate that at 10-MHz output and a 326.78-MHz clock frequency, the phase noise at a 19-kHz separation is about -135 dBc/Hz, which is within 3 dB of the clock phase noise.

Finally, it is important to ensure that the programming number is presented as a parallel data word with no time skew between bits. This will ensure that no spurious frequency occurs during a frequency change. A trace layout similar to that of Fig. 12.8 should help in this regard.

Chapter 13

Optimizing Universal Radio ICs

This chapter is devoted to optimizing ICs that are designed for a variety of radio or RF applications. The GEC Plessey SL6700 is selected as an example. The IC is designed for use in low-voltage AM applications. However, the versatility and access to internal functional blocks allow the SL6700 to be used in many more applications. The original use of the IC required the incorporation of a specialized noise blanker. The blanker circuit is still present and can be used, if desired. The low power consumption (less than 60 mW) of this IC makes it ideal for a number of applications. Design of the IC, which is optimized for use with ceramic filters, allows lower-cost circuitry to be obtained.

13.1 SL6700 Circuit Description

Figure 13.1 shows the functional circuits of the IC in block form. The following is a summary of these functions.

Amplifier 1, with an input on pin 18 and an output on pin 3, is AGC controlled.

Amplifier 2, with an input on pin 4 and an output on pin 6, is also AGC controlled, but has a lower signal-handling capacity than amplifier 1.

Double-balanced mixer, with an input on pin 7, local oscillator at pin 9, and an output at pin 8, has an open-collector output, and a third-order intermodulation intercept point of about −9 dBm.

Delay circuit, with an output at pin 5, provides a delayed AGC signal (the same AGC signal applied to amplifier 2).

AGC generator, with decoupling point at pin 16, provides an AGC signal to amplifier 2. This AGC signal is developed from the output of the amplifier/detector.

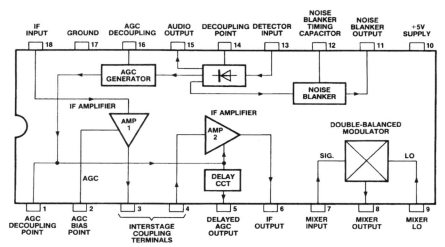

Figure 13.1 SL6700 block diagram (*GEC Plessey Semiconductors,* Professional Products IC Handbook, *p. 4-17*).

Amplifier/detector, with an input at pin 13, provides an audio output at pin 15, an internal signal to the AGC generator, and an internal signal to the noise blanker.

The noise blanker, with an output at pin 11, has a timing-capacitor connection point at pin 12.

13.2 SL6700 as a Double-Conversion IF Strip

Figure 13.2 shows the SL6700 connected as a double-conversion IF strip. The 10-MHz input is amplified in both IF amplifiers and converted to 455 kHz in the mixer. The 455-kHz output is filtered in a ceramic filter and applied to the detector input. AGC is applied to the two amplifier stages, and the variable resistor VR1, which sets the delayed AGC threshold. This AGC output is positive-going with increasing signal.

The sensitivity of this circuit is typically 5-μV RMS, with 30% modulation for a 10-dB signal-to-noise ratio, and will accept signals up to 100-mV RMS at 80% modulation, with distortion below 5%. The frequency response of the AGC-controlled amplifiers extends up to 25 MHz, allowing a wide choice of IF to suit any particular requirement.

Figure 13.3 shows the AM detector circuit. This type of detector produces both audio output and carrier-derived AGC voltages. The detector is full-wave with one input signal applied to an emitter-coupled transistor pair (TR53/TR54), which have common collectors. The input

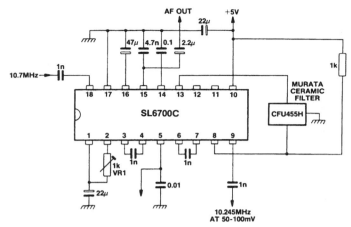

Figure 13.2 SL6700 as a double-conversion IF strip (*GEC Plessey Semiconductors,* Professional Products IC Handbook, *p. 4-17*).

signal is also phase-inverted by TR50/TR52 and applied to the other input of the detector. TR53 and TR54 act as a full-wave rectifier, and the emitters rise to the voltage determined by the input. The modulation also appears on the emitters, and is fed out at pin 15 along with the AGC voltage. The detector is linear. For example, the increase in audio output from 30% to 80% modulation is 8 dB, compared to a theoretical rise of 8.52 dB.

13.3 SL6700 as an AM Broadcast Radio

Figure 13.4 shows the SL6700 connected as an AM broadcast radio. The IC is ideally suited for this application because its high linearity and low distortion allow quality reproduction of AM broadcast signals. Also, the minimum of external components allows small size and low cost. For example, oscillator TR1 can be almost any small-signal NPN transistor. Standard broadcast-band components (ferrite loop sticks, etc.) and values can be used for C_1L_1 and C_2L_2. The calculations for such components are described in Chapters 1 and 2.

Filter F2 can be replaced with a 100-pF capacitor if a minimum component count is necessary, and degradation of selectivity is acceptable. The input of the mixer is typically 300 Ω, allowing a tight coupling to the coil (L1) without excessive loss of Q. This is an important consideration when using ferrite rod antennas.

The delayed AGC line is not used in this application, so it is left disconnected. If the circuit is to be used for shortwave broadcast reception, where an external RF amplifier is connected between the

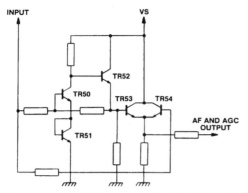

Figure 13.3 AM detector circuit of SL6700 (*GEC Plessey Semiconductors,* Professional Products IC Handbook, *p. 4-18*).

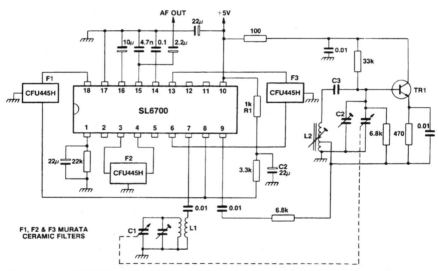

Figure 13.4 SL6700 as an AM broadcast radio (*GEC Plessey Semiconductors,* Professional Products IC Handbook, *p. 4-19*).

antenna and the SL6700 to improve sensitivity, and where an extra tuned circuit is used to increase image rejection, the 22-kΩ resistor from pin 1 to ground should be omitted. Instead, a 1-kΩ pot should be connected between pin 1 and ground (as in Fig. 13.2). In most cases, a resistance of 330 Ω between pin 1 and ground will provide suitable AGC characteristics. However, a variable resistance will provide for variation among ICs. The low supply voltage and current requirements make the circuit of Fig. 13.4 ideal for portable applications.

13.4 SL6700 as an AM/SSB/CW IF Strip

Figure 13.5 shows the SL6700 connected as an AM/SSB/CW IF strip. This is one of the more complex applications of the IC. The two gain-controlled amplifiers are cascaded and feed the detector and AGC circuits through a band-limiting filter. The amplifier output is also applied to the mixer input, which is used as a product detector. Although a ceramic filter is shown in Fig. 13.5, a suitable tuned circuit can be used. The load appearing across the circuit is low (about 1.5 kΩ). To get a suitable Q, it might be necessary to tap down the tuned circuit—either by means of capacitive or inductive taps. Because of the limited frequency response of the detector and the amplifier feeding the detector, the circuit is not recommended for use above about 1.6 MHz.

The audio output from the product detector is amplified in TR1 and the output is applied to the input of an SL621 AGC generator. The output of this stage controls the internal AGC circuitry through TR2. For AM operation, the carrier-insertion oscillator (BFO for CW operation) is switched off. As a result, there is little output from the product detector. The AM detector and AGC work normally, but the SL621 provides no output. In the SSB/CW modes, the product-detector output activates the SL621, and the AGC is taken over by this stage. However, RF-derived AGC is still applied in the event of strong signals at zero beat. R1 was chosen to set the outputs on AM and SSB approximately equal for SSB and 80%-modulated AM signals, respectively.

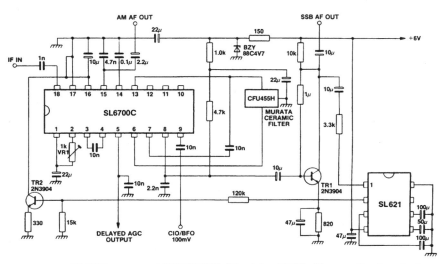

Figure 13.5 SL6700 as an AM/SSB/CW IF strip (*GEC Plessey Semiconductors, Professional Products IC Handbook, p. 4-19*).

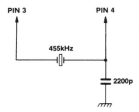

Figure 13.6 CW selectivity network (*GEC Plessey Semiconductors,* Professional Products IC Handbook, p. 4-20).

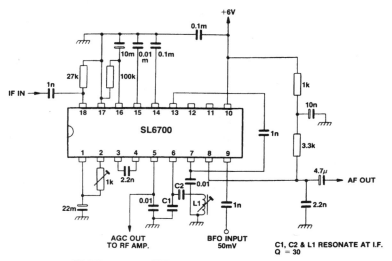

Figure 13.7 SL6700 as a CW receiver (*GEC Plessey Semiconductors,* Professional Products IC Handbook, p. 4-24).

Figure 13.6 shows a means to increase selectivity for CW reception. The 10-nF capacitor shown in Fig. 13.5 is replaced by a 455-kHz crystal in Fig. 13.6. This produces a very narrow peak to the passband. At 1.4 MHz or 1.6 MHz, the 2200-pF capacitor should be reduced to about 470 pF. The response will be less sharp at 455 kHz, but will still be adequately narrow for CW reception. The AM and SSB outputs should be kept completely separate and switch selected. During AM reception, the BFO/CIO must be switched off to avoid heterodyne interference and audio-derived AGC. If only CW reception is required, the circuit of Fig. 13.7 can be used, with a minimum of components.

The SL621C shown in Fig. 13.6 is a fast-attack SSB AGC generator. Because of the large capacitors used, the SL621C draws a large transient current from the supply when an SSB signal is first applied. To avoid instability, supply pin 4 of the SL621C should be decoupled with a 47-µF capacitor, using short leads. This will keep the ESR (equivalent series resistance) of the capacitor leads at a minimum.

The performance characteristics for the circuit of Fig. 13.5 are:

- Sensitivity in the AM mode is 7 dB SND/ND with a 5-µV RMS input and a modulation of 30%. With 80% modulation, the sensitivity is 15 dB SND/ND.
- Sensitivity in the SSB mode is 15 dB SND/ND with 30% modulation.
- AF output in the AM mode is 42 mV RMS with a 5-µV RMS input, 80% modulation, and a modulating frequency of 1 kHz.
- AF output in the SSB mode is 43 mV RMS with a modulating frequency of 1 kHz.
- AGC is 4 dB in the AM mode, and 5 dB in the SSB mode. This is a change in AF output for a change of 5 µV to 100 mV RMS input.
- Distortion in the AM mode is 2.8% with a V_{in} of 100 mV RMS, and 80% modulation at 1 kHz.
- Distortion in the SSB mode is 4.2% with a V_{in} of 100 mV RMS, and an f_{out} at 1 kHz.
- Signal-to-noise ratio in the AM mode is 28 dB with a V_{in} of 50 µV and 30% modulation. The S/N ratio increases to 36 dB with a V_{in} of 50 µV and 80% modulation.
- Signal-to-noise ratio in the SSB mode is 35 dB with a V_{in} of 50 µV and an f_{out} of 1 kHz.
- The ultimate S/N ratio is 50 dB in the AM mode and 40 dB in the SSB mode, with a V_{in} of 100 mV.

13.5 SL6700 as an SSB Generator

Figure 13.8 shows the SL6700 connected as an SSB generator. The SL6270 is a VOGAD (voice-operated gain-adjusting device), which maintains a constant output level for a wide range of inputs from the microphone. The input can be fed differentially, if desired, by removing the 2.2-µF capacitor from pin 5, and feeding the input between pins 4 and 5. The usual precautions should be taken against RF pickup in the transmitter. Bypass capacitors of 10 nF should be provided on the inputs.

The circuit of Fig. 13.8 has an advantage over the typical SSB-generator portion of an SSB transmitter in that no adjustment is required. Typically, the balanced-modulator portion of an SSB circuit requires an adjustment. This increases manufacturing costs

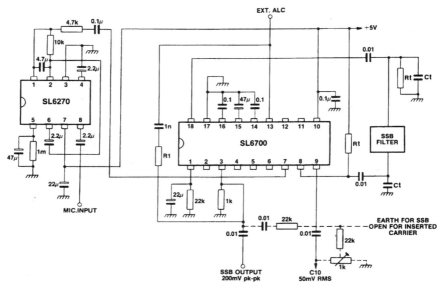

Figure 13.8 SL6700 as an SSB generator (*GEC Plessey Semiconductors,* Professional Products IC Handbook, *p. 4-22*).

and might cause problems after the equipment is placed in operation. In the Fig. 13.8 circuit, the mixer of the SL6700 is used as the balanced modulator, and provides about 20 dB of carrier suppression. For those not familiar with SSB transmitters and receivers, read *Lenk's RF Handbook* (McGraw-Hill, 1992).

The SSB filter provides additional carrier suppression (about 20 to 25 dB), resulting in about −40 dB relative to each tone. R_t and C_t are termination components for the SSB filter and should be chosen accordingly. However, impedances in the range of 1 to 4 kΩ are preferred. Less than 1 kΩ provides greater loss in the balanced modulator, thus degrading carrier suppression. Impedances greater than 4 kΩ usually require some special circuits to maintain the dc feed to the mixer.

For operation at 455 kHz, the recommended SSB filter is a Collins 526-9939-010, with an R_t of 2.7 kΩ and a C_t of 360 pF. For 1.4-MHz operation, the recommended filter is a Cathodeon BP4707/BP4708, with an R_t of 1 kΩ and a C_t of 15 pF. At 1.4 MHz, stray capacitances might make carrier leak somewhat worse. For that reason, it might be useful to connect a 6.8-kΩ resistor in series with pin 18 of the SL6700. This breaks a ground loop through the filter, and minimizes degradation of the carrier suppression.

Resistor R1 controls the input to the detector stage. In turn, this sets the AGC produced and the gain of the first amplifier. The value of R_1 also sets the ALC (automatic level control) threshold. Typical values for R1 are 47 kΩ at 1.4 MHz and 120 kΩ at 455 kHz, depending on the desired output and the amount of ALC required. An additional ALC input is available at pin 13. This input can be fed with an ALC voltage from a later stage in the transmitter (through a resistor) to provide multiple-level ALC action.

The 47-pF capacitor from pin 15 sets the ALC time constant, and should not be reduced. A reduction below 47 pF might result in distortion. The 1-kΩ resistor from pin 3 to ground increases the current through the emitter-follower driving this pin, and allows an undistorted output to be obtained with an impedance as low as 50 Ω.

For use in an SSB transceiver, the switching required between receive and transmit is probably too complex to be economical. The use of two SL6700s with a switched filter is recommended.

The performance characteristics for the circuit of Fig. 13.8 are:

- Carrier suppression is 50 dB at 455 kHz and 46 dB at 1.4 MHz, based on relative PEP (peak envelope power).
- Third-order IMD (intermodulation distortion) is −40 dB, relative for each tone of a two-tone signal, with separations down to 50 Hz.
- Second-order IMD is 43 dB at 455 kHz and 38 dB at 1.4 MHz.
- Output level is 200 mV p-p into a 600-Ω load.
- Carrier level is 50 mV RMS.
- Audio level is 30 mV RMS.

The circuit of Fig. 13.8 can be modified to provide for carrier reinsertion. This is shown by the dotted components. These components are connected to ground through a carrier switch or are left disconnected from ground. This arrangement provides for A1A, J2H, J3A, or J3H operation. The level of the carrier is set by the 1-kΩ pot. Use care when designing the carrier-switch return line to avoid ground-loop currents. For J3H operation, it is recommended that the carrier be set between −4 and −5 dB (relative PEP) to avoid distortion at modulation peaks. To minimize carrier "pumping" (where the carrier level depends on AF level) use no more than 2 or 3 dB of ALC.

Special care must be taken when the circuit of Fig. 13.8 is used above 1.6 MHz. This is because the balanced-modulator balance degrades as the frequency increases and the ALC detector sensitivity falls. However, the circuit has been used at frequencies up to 12 MHz with careful design and layout. Detector sensitivity can be increased by decreasing the R_1. Another problem at higher frequencies is that filters often do not provide the carrier separation that is obtainable at lower frequencies.

Figure 13.9 shows the SL6700C used as an SSB generator without the additional SL6270 VOGAD IC. This simple circuit does not have the input range of the Fig. 13.8 circuit, but it requires less power to operate (about 50 mW). The SL6270 is replaced by the second amplifier in the SL6700C.

13.6 SL6700 as a Remote-Control Receiver

Figure 13.10 shows the SL6700C connected as a remote-control receiver. The values listed are for a model radio-control (RC) receiver operating in the 27-MHz frequency range. The circuit will operate satisfactorily at 4.5 V with a current consumption of about 5 mA (for the SL6700). The supply voltage is best fixed at 4.5 to 5 V, with 6 V being an absolute maximum. The supply should never exceed 7 V, even momentarily, or permanent damage might result.

The component count can be minimized by replacing L1, the 56-pF capacitor, and the 1-nF capacitor with a ceramic filter, such as the Murata SFE 27MA4, and eliminating resistor R1 (connect pin 5 directly to ground).

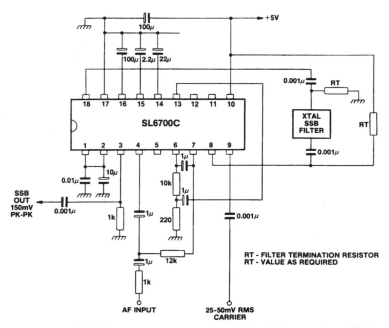

Figure 13.9 SL6700 as a minimum-component SSB generator (*GEC Plessey Semiconductors,* Professional Products IC Handbook, *p. 4-23*).

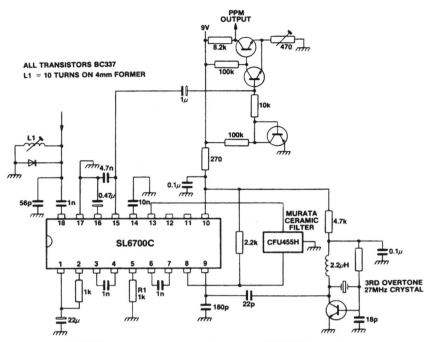

Figure 13.10 SL6700 as a remote-control receiver (*GEC Plessey Semiconductors, Professional Products IC Handbook, p. 4-23*).

Chapter 14

Optimizing Log/Linear Amplifier ICs

This chapter is devoted to optimizing log/linear amplifier ICs. The GEC Plessey SL3522 is selected as an example. This IC is a complete successive-detection log/limiting amplifier that can operate over an input frequency range of 100 MHz to 650 MHz. The amplifier produces a log/linear characteristic for signals between −68 dBm to +7 dBm, with an accuracy of ±1 dB.

14.1 SL3522 Circuit Description

Figures 14.1 and 14.2 show the internal circuits and typical application/test circuit, respectively, for the SL3522. The amplifier consists of six gain stages, seven detector stages, a limiting RF-output buffer, and a video-output amplifier. The power-supply connections to each section are isolated from each other to improve stability. Each of the gain stages and detector stages has about 13 dB of gain and a significant amount of on-chip RF decoupling (which also aids stability).

The limiting RF-output buffer provides a balanced limited output of 0 dBm on each RF-output line (pins 9 and 10), for input signal levels in excess of −65 dBm (when input is applied to pins 27 and 28). The output buffer can be isolated from the other parts of the log/limiting amplifier by disconnecting the output buffer ground (pin 8) from 0 V. This feature improves stability in applications that do not require a limited RF output.

The video amplifier provides a positive-going output signal that is proportional to the log of the RF-input amplitude at pins 27 and 28. This is shown in Fig. 14.3. The gain and offset of the video amplifier

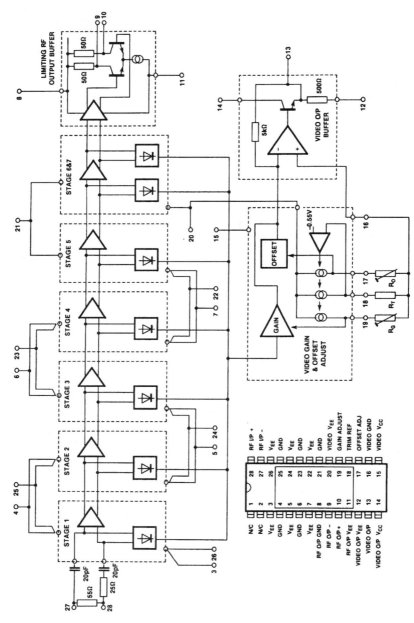

Figure 14.1 SL3522 schematic diagram (*GEC Plessey Semiconductors*, Professional Products IC Handbook, *p. 4-76*).

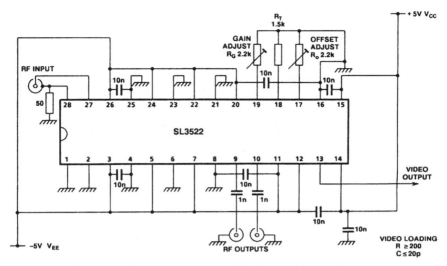

Figure 14.2 SL3522 application/test circuit (*GEC Plessey Semiconductors,* Professional Products IC Handbook, p. 4-73).

can be adjusted by three resistors: R_G, R_T, and R_O. Resistor R_G at pin 19 provides for gain adjustment. Resistor R_T at pin 18 provides a trim reference. Resistor R_O at pin 17 provides an offset adjustment. With R_T set to 1.5 kΩ, R_G can be set to any value between 1 kΩ and 2.2 kΩ to achieve a range in log slope of ±20%, centered on 20 mV/dB. Resistance R_O can be set to any value between 1 kΩ and 2.2 kΩ to provide an offset voltage in the range between −0.5 V and +1.0 V.

The RF-input pins (27 and 28) have a 50-Ω terminating resistor connected between them on the chip. In turn, the pins are capacitively coupled to the input gain stage (stage 1). The RF input can be driven either balanced or unbalanced (single ended).

The IC consumes about 1.1 W of power when all of the internal circuits are used. If the RF output buffer is not used (pin 8 left floating), the power consumption drops to about 0.95 W. In both cases, the IC should be powered from a ±5-V supply. Because the stages are operated as differential class A, the gain stages, detectors, and RF-output buffers are independent of the input signal level. However, the video output at pin 13 is single ended. As a result, the power consumption of the video amplifier will vary with the RF-input signal level on pins 27 and 28.

14.2 Optimizing the RF-Output Buffer

The IC has about 70 dB of broadband RF gain, with 3 dB of bandwidth, at 450 MHz. This gain is available at pins 9 and 10 when the RF-output

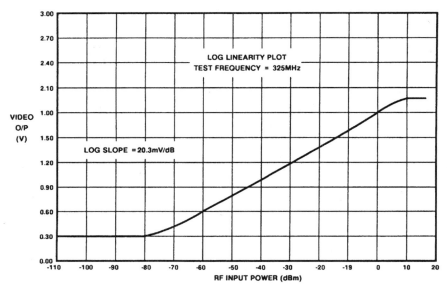

Figure 14.3 Typical log/linear characteristics for the SL3522 (*GEC Plessey Semiconductors,* Professional Products IC Handbook, *p. 4-74*).

buffer is used (pin 8 connected to ground). Operating the IC with the RF-output buffer off (pin 8 floating) removes the RF output, and results in a log amplifier with an output at pin 13 (video output). When the IC is used only as a log amplifier, the stability is increased and the power consumption is decreased.

Pin 8 is the positive supply connection for the RF-output buffer. If pin 8 is open, power consumption is decreased by about 30 mA. The V_{EE} pin (11) for the RF-output buffer should always be connected to the V_{EE} supply—even if the RF-output buffer is not used.

14.3 Optimizing SL3522 Circuits

Observe the following precautions when using the IC for any application.

Mount the IC on a ground plane. Use RF-quality chip capacitors for all supply decoupling. Keep the leads to the decoupling capacitors as short as possible.

Connect the RF V_{EE} pins (3, 5, 7, 20, 22, 24, and 26) to a low-impedance copper plane. Use a two-layer PC board, if possible.

Return the load current at the video-output pin (13) to the video-output V_{CC} pin (14) through a 10-nF capacitor. Connect the capacitor to the video load return line to avoid any common-impedance path.

Decouple the video-output V_{EE} pin (12) directly to the video-output V_{CC} pin (14) with a 10-nF capacitor.

Optimizing Log/Linear Amplifier ICs 211

The unscreened lead length at the RF input should be kept to a bare minimum. If the input source is single ended, connect the RF-input return line to the ground plane through a 50-Ω chip or bead resistor at pin 28. Keep the RF return line very close to the ground plane. (The RF input can also be driven differentially.) Figure 14.4 shows the Smith chart representation of the RF input, normalized to 50 Ω, with a −20-dBm input level.

As shown in Fig. 14.1, the RF input is isolated from the input amplifier (stage 1) by two series on-chip capacitors. The IC also includes an on-chip termination resistor, connected directly across the RF input pins (27 and 28).

If the IC is operated with the RF-output buffer powered up (pin 8 at ground), take care to present both the RF-output pins (9 and 10) with matched loads. Ideally, each pin should be loaded with a 50-Ω terminated transmission line. The IC stability is very sensitive to imbalance at the output. Driving highly reactive loads (with standing waves) is not recommended (for stability reasons).

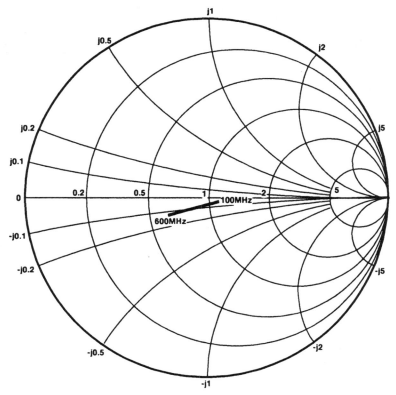

Figure 14.4 Smith chart representation of SL3522 RF input (*GEC Plessey Semiconductors,* Professional Products IC Handbook, *p. 4-75*).

Although the RF-output pins (9 and 10) have dc-blocking capacitors as shown in Fig. 14.2, these pins can be operated with a dc load to ground. However, a dc offset of about 400 mV will exist on each pin. Also, it is not possible to power-down the RF-output buffer when pins 9 and 10 are to be used in any configuration.

14.4 Optimizing Video Performance

To get the best video transient response, the trim-reference resistor, R_T (at pin 18) must have a parasitic capacitance of less than 5 pF. Typically, the value of R_T is 1.5 kΩ. The load at the video output must not be less than 200 Ω resistive and more than 20 pF capacitive.

14.5 Gain and Offset Trimming

Gain and offset trimming are unilaterally independent. That is, adjustment of gain (R_G) has an effect on the offset (R_O), but adjustment of the offset does not affect gain. Gain and offset control depend on the difference between the R_G (or R_O) resistance values and the value of trim-reference resistor R_T, but not directly on the value of R_G or R_O. As a result, all three resistors should have the same temperature coefficients. If not, there will be variations in gain and/or offset with changes in temperature. It is recommended that resistors of an identical type be used, preferably from the same batch.

Index

A

A counter, 148–149, 181–182
Admittances:
  adding, 40–41
  transforming, 40
AFT (*see* Automatic fine-tuning)
Air-core coils, inductance of, 29
AM broadcast band, 2
AM section (of IC RF voltage amplifier), 13–14
AM spectrum displays, 73–76
AM tuning (of IC RF voltage amplifier), 15–18
Amplifiers, RF, 1–23
  bandwidth of, 3
  efficiency of, 52
  examples of circuits for, 6–8
  gain calculations for, 44–45
  general-purpose, 117–122
  IC design, 12–23
  log/linear amplifier ICs (*see* Log/linear amplifier ICs)
  mismatching, 43–44, 46
  multipliers vs., 47–48
  narrowband, 3–6
  neutralization of, 43–46
  stability of, 41, 43–46
  tuned amplifier, 124–127
  types of, 1–2
  wideband, 8–12
Antenna booster, wideband VHF, 123–124
Automatic fine-tuning (AFT), 16, 19

B

Bands, RF, 1
Bandwidth, 3
Bessel function, 76

Bias:
  in discrete-component RF-oscillator circuit design, 60–61
  in large-signal design approach, 49
Broadband RF amplifiers (*see* Wideband RF amplifiers)
Bypass capacitors:
  in discrete-component RF-oscillator circuit design, 63
  in large-signal design approach, 50

C

Capacitance:
  collector-to-base, 45
  coupling, 10
  distributed, 9
Capacitance-bridge technique, 45, 46
Capacitors:
  bypass, 50, 63
  coupling, 63
  midrange, 48
  parallel, 48
Carrier frequency, 75
Cascading, 8
CMOS outputs, high-speed dividers with, 166–167
Coarse error signal, 177
Coil inductance measurements, 78–80
Collector leakage, 66
Collector resistance, 10
Collector-to-base capacitance, 45
Common-base RF circuits, 6
Constant resistance (in Smith chart), 38
Coupling capacitance, 10
Coupling capacitors, 63
Crystal-control oscillators, 58–59, 64–65

214 Index

D

Damping resistances, 11–12
DBS, tuners for digital (*see* Digital direct-broadcast satellite tuners)
Design, RF-circuit, 30–31
   discrete-component RF-oscillator, 57–65
   large-signal approach, 46–57
   Smith charts, using, 38–42
   and stability, 41, 43–46
   thermal design considerations, 65–68
   y-parameters approach, 31–38
DFSs (*see* Direct frequency synthesizers)
Digital direct-broadcast satellite (digital DBS) tuners, 85–93
   baseband amplifiers, 90
   circuit description for, 85–87
   external oscillator requirements for, 89
   filter requirements for, 91–93
   front-end tuner circuitry for, 87–89
   layout optimization for, 91
   offset correction for, 90
   power-supply sequencing with, 91
   prescaler requirements for, 89–90
Digital tuning (*see* Frequency synthesis tuning)
Diode ring, 131
Direct frequency synthesizers (DFSs), 185–193
   circuit in, 185–190
   optimization of, 190–193
   spurious outputs, minimization of, 192–193
Discrete-component RF-oscillator circuit design, 57–65
   bias-circuit components, 60–61
   bypass capacitors, 63
   and C values, 60
   coupling capacitors, 63
   and crystal frequency, 60, 63
   crystal-control oscillators, 58–59, 64–65
   and feedback, 60, 61
   and frequency, 61–62
   and frequency stability, 60
   LC oscillators, 57–59
   output circuit, 63
   resonant circuits, L/C combinations for, 62–63
   RF chokes, 63
Distributed capacitance, 9, 80
Double-conversion PLL detector/RF mixer, 135–142
   IF amplifier/mixer, optimization of, 135–137
   loop filter, optimization of, 139–142

Double-conversion PLL detector/RF mixer (*Cont.*):
   outputs with, 138
   PLL, optimization of, 137
   squelch, optimization of, 138
   VCO adjustment with, 138–139
   VCO frequency grading, 138
Dynamic range (DR), 127–128

E

ECL outputs, high-speed dividers with, 166
Electromagnetic waves, 1
Emitter bypass, 10
Error signal, 18
Error voltage, 15
Extended PLL, 18

F

Feedback:
   in discrete-component RF-oscillator circuit design, 60, 61
   inverse, 12
   with narrowband RF amplifiers, 4–6
   negative, 12
Ferrite beads, 50
Field-effect transistors (FETs), 4–5, 32–35, 37, 127, 130–131
Filter(s):
   digital DBS tuners, 91–93
   double-conversion PLL detector/RF mixer, 139–142
   frequency synthesis circuits, 156–162
   IF transceiver with limiter and RSSI, 103–104
Fine error signal, 177
FM broadcast band, 2
FM section (of IC RF voltage amplifier), 12–13
FM spectrum displays, 74, 76–77
FM tuning (of IC RF voltage amplifier), 15, 18
Forward transadmittance (y_{fs}), 34, 35
Fourier analysis, 73
Frequency selection, 7
Frequency synthesis (FS) circuits, 143–162
   A/M divider programming, 181–182
   basic single-loop PLL circuit, 175, 176
   calculator program, 182–183
   characteristics of, 148
   data entry/storage with, 151–154
   external mode, addressing in, 177

Index

Frequency synthesis (FS) circuits (*Cont.*):
  IC pin functions, 177–179
  and loop bandwidth, 143–144
  loop filter, optimization of, 156–162
  and multimodulus division, 145–148
  optimization of, 154–157, 175–183
  phase comparator in, 150–151
  phase comparators with, 179–180
  prescaler/A/M counters in, 148–149
  programmable reference divider in, 150
  programming considerations with, 179
  reference divider, 180
  reference source, obtaining, 149–150
  self-programming internal mode, addressing in, 175–177
  single-shot internal mode, addressing in, 177
Frequency synthesis (FS) tuning, 16–22
  AM FS tuning, 16, 18
  and elements of FS system, 16, 17
  FM FS tuning, 18
  TV FS tuning, 19–22
FS circuits (*see* Frequency synthesis circuits)
FS tuning (*see* Frequency synthesis tuning)

G

Gain:
  in large-signal design approach, 52
  log/linear amplifier ICs, 212
  maximum available, 44–45
  maximum usable, 44, 45
  optimization of tuned amplifier for maximum, 126–127
  and temperature, 66
  unilaterlized, 44
General-purpose RF amplifiers, 117–122
  circuit characteristics, 117–118
  external components, 118
  layout optimization for, 119–122
Grounded emitter (large-signal design approach), 49
Grounding, single-point, 164

H

High-speed frequency-divider ICs, 163–173
  CMOS outputs, dividers with, 166–167
  components for, 163–164
  ECL outputs, dividers with, 166
  ECL-TTL interface circuit, 167

High-speed frequency-divider ICs (*Cont.*):
  frequency range/input level with, 165–166
  impedance matching with, 164–165, 167–168
  interfacing, 167
  in microwave synthesizers, 172
  open-collector TCL outputs, dividers with, 166
  PC boards for, 163
  single-point grounding with, 164
  true TTL outputs, dividers with, 166
  variable-modulus dividers, optimization of, 168–171
  in VHF synthesizers, 172–173
Hybrid notation systems, 34

I

IC RF voltage amplifiers, 12–23
  AM section of, 13–15
  AM tuning in, 15–16
  FM section of, 12–13
  FM tuning in, 15
  frequency tuning of, 16–18
  frequency-synthesis tuning of, 19–22
  tuner circuits, relationship of, 12–16
  TV IF/video-detector units, 22–23
ICs:
  calculation of power dissipation for, 67–68
  log/linear amplifier (*see* Log/linear amplifier ICs)
  universal radio (*see* Universal radio ICs)
IF (*see* Intermediate frequency)
IF transceiver with limiter and RSSI, 95–105
  200- to 440-MHz RF applications, 101
  circuit description for, 95–96
  filter sharing in, 103–104
  image-rejection mixer in, 96–97
  impedance matching for input, 102–103
  layout optimization for, 104–105
  limiter in, 98–99
  local-oscillator buffer, 100
  oscillator tank calculations for, 101–102
  power-supply modes, 100–101
  receiver circuit in, 96–99
  transmitter circuit in, 99–100
Imaginary part, 8
IMD (*see* Intermodulation distortion)
Impedance matching, 7–8, 164–165, 167–168

216 Index

Impedance measurements, resonant-circuit, 82–83
Impedances:
  adding, 40–41
  transforming, 40
Incidental FM, 75
Inductance:
  calculation of, 28–29
  of RF coils, 29
  variable, 25
Input admittance (y_{is}), 33–34, 36
Intermediate frequency (IF), 22–23
Intermodulation distortion (IMD), 127–130
Interstage coupling network, 7
Inverse feedback, 12

L

Large-signal characteristics, 47
Large-signal design approach, 46–57
  and amplifier efficiency, 52
  and bias, 49
  bypass capacitors in, 50
  calculations, 53–55
  and efficiency of intermediate amplifiers, 52
  ferrite beads in, 50
  grounded emitter in, 49
  microstrip circuits, 50–51
  midrange capacitors in, 48
  multiplier circuits, 47–48
  and network characteristics, 55–56
  parallel capacitors in, 48
  parameters in, 53, 54
  and power gain, 52
  resonant network, 53
  RFC connections in, 49
  and RFC ratings, 49–50
  and transistor characteristics, 52
  tuning controls in, 48
Loading control, 48
Local-oscillator (LO) buffer, 85, 87–91, 100
Log/linear amplifier ICs, 207–212
  description of, 207–209
  gain/offset trimming with, 212
  optimization of, 209–212
Loop filter optimization:
  in double-conversion PLL detector/RF mixer, 139–142
  in frequency synthesizers, 156–162
  second-order calculations, 157–159
  third-order calculations, 159–162

M

M counter, 148–149, 181–182
MAG (*see* Maximum available gain)
MAX2102, 85–93
  baseband amplifiers, 90
  circuit description, 85–87
  external oscillator requirements, 89
  filter requirements, 91–93
  front-end tuner circuitry, 87–89
  layout, optimization of, 91
  offset correction, 90
  power-supply sequencing with, 91
  prescaler requirements, 89–90
MAX2511, 95–105
  200- to 440-MHz RF applications, 101
  circuit description, 95–96
  filter sharing, 103–104
  image-rejection mixer, 96–97
  impedance matching for input, 102–103
  layout, optimization of, 104–105
  limiter, 98–99
  local-oscillator buffer, 100
  oscillator tank calculations, 101–102
  power-supply modes, 100–101
  receiver circuit, 96–99
  transmitter circuit, 99–100
MAX2601/2602, 107–110
  circuit characteristics for, 107–108
  current-mirror bias for, 108–109
  layout, optimization of, 110
  optimum port impedance with, 109–110
MAX2620, 111–115
  circuit characteristics, 111–112
  output buffers, 112–115
  tank-circuit design, 113–114
MAX2630-MAX2633, 117–122
  circuit characteristics, 117–118
  external components, 118
  layout, optimization of, 119–122
Maximum available gain (MAG), 44–45
Maximum frequency deviation, 76
Maximum power dissipation, 66
Maximum usable gain (MUG), 44, 45
Measurements, RF, 77–83
  coil inductance measurements, 78–80
  probes, RF, 77
  resonant-circuit impedance measurements, 82–83
  resonant-circuit Q measurements, 80–82
  resonant-frequency measurements, 77–78
  self-resonance and distributed-capacitance measurements, 80

Microstrip circuits, 50–51
Microwave synthesizers, 172
Microwaves, 1
Midrange capacitors (in large-signal design approach), 48
Miller effect, 4–5
Mismatch, 43–44, 46
Mixers, high-performance, 127–134
  circuit characteristics, 131–133
  diode ring, 131
  intermodulation distortion with, 127–130
  optimization of, 134
  quad FET mixers, 131
  single FET, 130–131
  single-diode mixer, 130
  thermal considerations with, 133–134
Modulation frequency, 75
Modulation percentage, 75
MUG (*see* Maximum usable gain)
Multimodulus division, 145–148
Multiplex decoder, 12
Multipliers, RF, 8, 47–48

N

Narrowband RF amplifiers, 3–6
  circuit examples, 6–8
  feedback problems with, 4–6
  tuning of, 3–4
Negative feedback, 12
Neutralization, 5–6, 31, 43, 45–46
NJ8820/21/23, 175–183
  A/M divider programming, 181–182
  basic single-loop PLL circuit, 175, 176
  calculator program, 182–183
  external mode, addressing in, 177
  IC pin functions, 177–179
  phase comparators with, 179–180
  programming considerations with, 179
  reference divider, 180
  self-programming internal mode, addressing in, 175–177
  single-shot internal mode, addressing in, 177
Nonlinear modulation, 75

O

Oscillators, double-conversion PLL detector/RF mixer, 137
Oscillators, IC, 111–115
  circuit characteristics for, 111–112
  output buffers, 112–115
  tank-circuit design, 113–114

Oscillators (crystal-control), 58–59, 64–65
Output admittance (y_{os}), 34, 36
Output buffers:
  MAX2620, 112–115
  SL3522, 209–210
Output power, 53
Output reactance, 53
Output resistance, 53
Overmodulation, 76

P

Parallel capacitors (in large-signal design approach), 48
PC boards, 163
Peaking coil:
  series, 11
  shunt, 10–11
Phase-locked loop (PLL), 15, 16, 18–21, 101
  double-conversion PLL detector (*see* Double-conversion PLL detector/RF mixer)
  frequency synthesizer optimization, 143–145, 175, 176
Power, output, 53
Power dissipation:
  for ICs, 67–68
  for SL6440, 133–134
  for transistors, 66–67
Power transistors for 900 MHz, 107–110
  circuit characteristics for, 107–108
  current-mirror bias for, 108–109
  layout optimization for, 110
  optimum port impedance with, 109–110
Prescaler, 18
Probes, RF, 77
Programmable divider, 18
PSC (*see* Pulse-swallow control)
Pulse amplifiers, 8
Pulse swallowing (*see* Multimodulus division)
Pulse-swallow control (PSC), 18, 19

Q

Q factor, 27–29
Q measurements, 80–82
Quad FET mixers, 131
Quartz tuning (*see* Frequency synthesis tuning)

R

Radio ICs, universal (*see* Universal radio ICs)

218 Index

Radio-frequency (RF) signals, 1
RC amplifiers (see Resistance-coupled amplifiers)
Reactance:
  output, 53
  in Smith chart, 38, 40
Real part, 8
Reference-frequency division ratio, 148
Reflection coefficient, 70
Resistance-coupled (RC) amplifiers, 8–9
Resistance(s):
  collector, 10
  damping, 11–12
  output, 53
  thermal, 66
Resonant circuits, 25–27
  equations for, 26
  large-signal design, 53–57
  selectivity of, 27
Resonant frequency:
  calculating, 28–29
  and Q factor, 27–28
Resonant networks, 53
Resonant-circuit impedance measurements, 82–83
Resonant-circuit Q measurements, 80–82
Resonant-frequency measurements, 77–78
Reverse transadmittance (y_{rs}), 34, 36–37
RF chokes (RFCs), 49–52, 63
RF (radio-frequency) signals, 1

S

Safe-operating-area curves, 66
Sample carrier signal, 22
Satellite, direct-broadcast (see Digital direct-broadcast satellite tuners)
Selectivity (of resonant circuit), 27
Self-resonance measurements, 80
Series peaking coil, 11
Shunt peaking coil, 10–11
Single FET, 130–131
Single-diode mixers, 130
Single-point grounding, 164
SL3522, 207–212
  circuit description, 207–209
  circuits, optimization of, 210–212
  gain/offset trimming with, 212
  RF-output-buffer, optimization of, 209–210
  video performance, optimization of, 212
SL6440, 127–134
  circuit characteristics, 131–133
  diode ring, 131

SL6440 (Cont.):
  intermodulation distortion with, 127–130
  optimization of, 134
  quad FET mixers, 131
  single FET, 130–131
  single-diode mixer, 130
  thermal considerations with, 133–134
SL6601, 135–142
  IF amplifier/mixer, optimization of, 135–137
  loop filter, optimization of, 139–142
  outputs with, 138
  PLL, optimization of, 137
  squelch, optimization of, 138
  VCO adjustment with, 138–139
  VCO frequency grading, 138
SL6700, 195–205
  as AM broadcast radio, 197–198
  as AM/SSB/CW IF strip, 199–201
  circuit description, 195–196
  as double-conversion IF strip, 196–197
  as remote-control receiver, 204–205
  as SSB generator, 201–204
Slug adjustments, 25
Smith chart, 38–42
  addition of impedances/admittances, 40–42
  construction of, 38–40
  transformation of impedances/admittances, 40
SP2001:
  circuit description, 185–190
  optimization of, 190–193
  spurious outputs, minimization of, 192–193
SP8853, 143–162
  characteristics of, 148
  data entry/storage with, 151–154
  and loop bandwidth, 143–144
  loop filter, optimization of, 156–162
  and multimodulus division, 145–148
  optimization of, 154–157
  phase comparator in, 150–151
  prescaler/A/M counters in, 148–149
  programmable reference divider in, 150
  reference source, obtaining, 149–150
Special-purpose RF ICs, 123–142
  amplifier, tuned, 124–127
  antenna booster, wideband VHF, 123–124
  double-conversion PLL detector/RF mixer, 135–142
  mixer, high-performance, 127–134
Spectrum analysis, 71–72

Index

Spectrum display(s), 72–73
  AM, 73–76
  FM, 74, 76–77
  unmodulated, 73
Squelch optimization, 138
Standing waves, 68
Standing-wave-ratio measurement, 68–71. (*See also* Voltage standing wave ratio)
Stern solution, 43
SWR (*see* Voltage standing wave ratio)

T

Test equipment, RF, 68–77
  AM spectrum displays, 73–76
  FM spectrum displays, 74, 76–77
  Fourier/transform analysis, 73
  spectrum analysis, 71–72
  spectrum display, 72–73
  standing-wave-ratio measurement, 68–71
  unmodulated spectrum displays, 73
Thermal design, 65–68
  and effects of temperature, 66
  ICs, calculation of power dissipation for, 67–68
  transistors, calculation of power dissipation for, 66–67
Thermal resistance, 66
Thermal runaway, 65
Transform analysis, 73
Transistors:
  calculation of power dissipation for discrete, 66–67
  in large-signal design approach, 52
  and Miller effect, 4–5
Transmitting stations, 2
TTL outputs, high-speed dividers with, 166
Tuned amplifier, 124–127
  circuit description, 125–126
  maximum gain, optimizing for, 126–127
Tuning controls (in large-signal design approach), 48
Tuning substitution, 35
Tuning-correction voltage, 18
TV:
  broadcast bands for, 2
  FS tuning circuit, 19, 21–22
  pulse-swallow control, 19

U

Unilaterlized gain, 44
Universal radio ICs, 195–205

Universal radio ICs (*Cont.*):
  as AM broadcast radios, 197–198
  as AM/SSB/CW IF strips, 199–201
  description of, 195–196
  as double-conversion IF strips, 196–197
  as remote-control receivers, 204–205
  as SSB generators, 201–204
Unmodulated spectrum displays, 73
Unstable circuits, 30–31

V

Variable-frequency oscillator (VFO), 16
Variable-modulus dividers, optimizing, 168–171
VCO, 15, 138–139, 143–145
VF (video-frequency) amplifiers, 8
VFO (*see* Variable-frequency oscillator)
VHF antenna booster, wideband, 123–124
VHF synthesizers, 172–173
Video-frequency (VF) amplifiers, 8
Voltage standing wave ratio (VSWR), 40, 69–70, 85
Voltage-variable capacitors (VVCs), 29–30
VSWR (*see* Voltage standing wave ratio)
VVCs (*see* Voltage-variable capacitors)

W

Wideband RF amplifiers, 8–12
  circuit for, 9
  collector resistance with, 10
  coupling capacitance with, 10
  damping resistances with, 11–12
  emitter bypass with, 10
  increasing wideband response with, 9–10
  inverse feedback with, 12
  series peaking coil in, 11
  shunt peaking coil in, 10–11
  types of, 8
Wideband VHF antenna booster, 123–124
Window detectors, 22

Y

y-parameters, 31–38
  forward transadmittance, 34
  input admittance, 33–34
  measuring, 34–38
  output admittance, 34
  reverse transadmittance, 34
  and vector algebra, 31–32

ABOUT THE AUTHOR

John D. Lenk has been a technical author specializing in practical electronic design/service guides for over 40 years. He is the best-selling author of more than 90 books on circuit and consumer electronics, which together have sold over 2 million copies in nine languages and 33 countries. His most recent titles include *Lenk's Video Handbook, Lenk's Digital Handbook, Lenk's Audio Handbook, Lenk's Laser Handbook, Lenk's RF Handbook, Lenk's Television Handbook, the McGraw-Hill Circuit Encyclopedia,* Volumes 1-4, the *McGraw-Hill Electronic Testing Handbook,* and the *McGraw-Hill Electronic Troubleshooting Handbook.*